150 Best New House Ideas

150 Best New House Ideas

COLLINS DESIGN

An Imprint of HarperCollins Publishers

150 BEST NEW HOUSE IDEAS
Copyright © 2008 by COLLINS DESIGN and LOFT Publications

HarperCollins books may be purchased for educational, business, or sales promotional use.
For information, please write: Special Markets Department, HarperCollins*Publishers*,
10 East 53rd Street, New York, NY 10022.

First published in 2008 by:
Collins Design
An Imprint of HarperCollins*Publishers*
10 East 53rd Street
New York, NY 10022
Tel.: (212) 207-7000
Fax: (212) 207-7654
collinsdesign@harpercollins.com
www.harpercollins.com

Distributed throughout the world by:
HarperCollins*Publishers*
10 East 53rd Street
New York, NY 10022
Fax: (212) 207-7654

Executive editor:
Paco Asensio

Editorial coordination:
Catherine Collin

Editor and texts:
Bridget Vranckx

Art director:
Mireia Casanovas Soley

Cover design:
Claudia Martínez Alonso

Layout:
Ignasi Gracia Blanco
María Eugenia Castell Carballo

Library of Congress Cataloging-in-Publication Data

Vranckx, Bridget.
  150 best new house ideas / by Bridget Vranckx.
    p. cm.
  ISBN 978-0-06-153792-9
  1. Architect-designed houses. 2. Architecture, modern--21st century. I. Title. II. Title: One hundred fifty
best new house ideas.

  NA7125.V73 2008
  724'.6--dc22

                              2008011638

ISBN: 978-0-06-153792-9

Printed in China
First Printing, 2008

Contents

Introduction

Twenty-first century house design has come a long way since our forefathers' first attempts at putting a roof over their heads. Today and tomorrow's designs are a far cry from the adobe structures of ancient tribes and house trends have evolved over the centuries as new materials and technologies reshape the way we build. Styles vary from country to country and person to person as each culture and individual has different needs and possibilities. *150 Best New House Ideas* complements the first volume in this series and presents a variety of home designs from around the world.

Despite the shortage of space and the unaffordable housing prices, urbanites the world over are determined to set up home in some of the world's largest metropolises and to sacrifice space over location, creating the architect's ultimate challenge. Chapter one will lend some ideas on how to build a comfortable home in the city in spite of site constraints. Other site difficulties such as unusual slopes, hills, floodplains, a harsh surrounding landscape, or a difficult-to-reach plot are the focus of the second chapter, with examples in the United States and further afield in Switzerland, Japan, Argentina and Australia.

As in other areas, increased sensitivity to the environment has also reached contemporary home design and is possibly one of this field's most important current trends. Chapter three offers more than twenty tips to help build an ecologically sound home, whether completely or partially. If you are looking to extend or remodel your current home, turn to chapter four for some examples. Meanwhile, seemingly impossible odd requests, unusual site constraints or budget limitations all find their unique solution in chapter five, for example, how to incorporate changing lifestyles in living spaces, or comply with view restrictions, without compromising the inhabitant's comfort. And, finally, to add a touch of the spectacular to your home, or help create a fabulous dwelling of your own, turn to page 508.

Urban

Sliver House

A unique response to a difficult site, Sliver House is built on a narrow infill site between two large Victorian end-of-terrace buildings. The architects were not only faced with the challenge of how to fit a house in a space directly overlooked by several windows, but also with the problem of bringing as much light as possible into the house without sacrificing privacy and respecting the conservation area.

Architect: Boyarsky Murphy
Architects
Location: Elgin Avenue, London,
UK
Date of construction: 2006
Photography: Helene Binet

The master bedroom floor
in the rear elevation of the
house is cantilevered
over the lower floors towards
the sun.

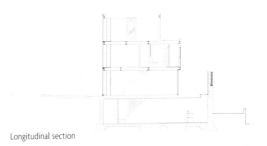

Longitudinal section

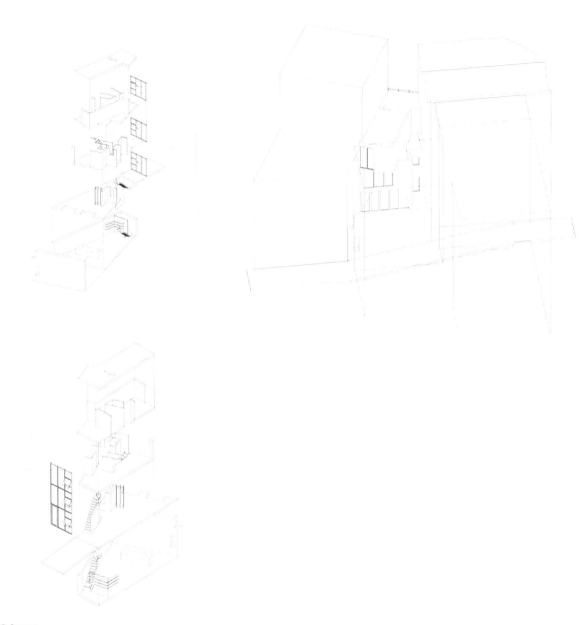

3D diagrams

1. Kitchen
2. Dining room
3. Utility
4. Patio
5. Roof lights
6. Living room
7. Entry

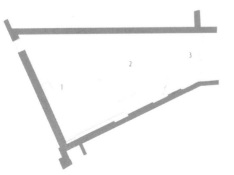

Basement

First floor

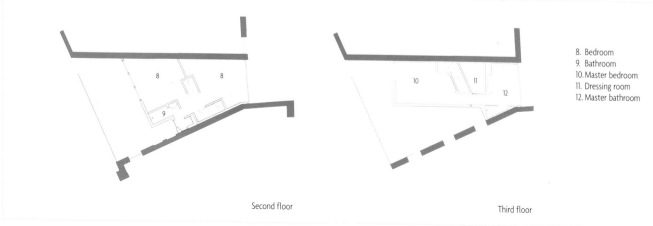

8. Bedroom
9. Bathroom
10. Master bedroom
11. Dressing room
12. Master bathroom

Second floor

Third floor

House SH

Situated in a densely built residential area of Tokyo on a 420 square-foot lot, this single-family house is only open on the north-facing street side. The clients wanted a light house, but also privacy and security. The architects play with the wall, pushing it out to the maximum building coverage. This wall catches the light which filters down from a large skylight and spreads over the three floors.

Architect: Hiroshi Nakamura/
NAP Architects
Location: Tokyo, Japan
Date of construction: 2005
Photography: Daici Ano

Site plan

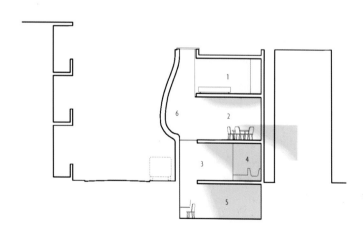

1. Children's room
2. Living/dining room
3. Entrance hall
4. Bathroom
5. Bedroom
6. Light well

North-south section

N
S=1/150

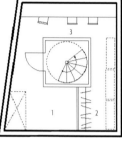

First floor

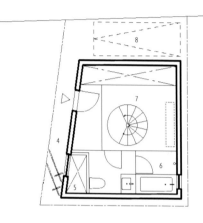

Second floor

1. Bedroom
2. Closet
3. Reading room
4. Bicycle parking
5. Light well
6. Bathroom
7. Entrance hall
8. Parking
9. Living room
10. Kitchen
11. Dining room
12. Children's room

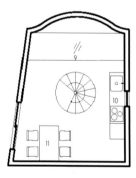

Third floor

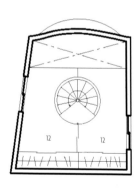

Fourth floor

A soft hollow in the wall
catches the sunlight from the
skylight above and distributes
it to the other levels creating
different expressions of light
for each floor.

The wall will become a part of
the inhabitant's lives,
communicating with them and
serving them as a comfortable
bench to sit or lie on.

The children's rooms are at the top of this three-storey house. A lightwell helps distribute the light to the rest of the house below.

Pachter Studio

Charles Pachter decided to turn the small parking lot situated in front of his Moose Factory studio into a gallery-cum-studio-cum-living quarters. The project, which has earned numerous design awards, is designed around three "tubes." Each "tube" contains an aspect of the artist's life—studio, gallery, residence. Its extreme transparency turns the artist's daily life into a work of art.

Architect: Teeple Architects
Location: Toronto, Canada
Date of construction: 2005
Photography: Tom Arban

This unique home comfortably fits a gallery, studio and residence into an 18-by-100-foot parking lot by organizing the program of the house vertically.

1. Front garden
2. Entrance walkway
3. Studio workroom
4. Pool/courtyard
5. Gallery
6. Courtyard
7. Back studio
8. Living room
9. Kitchen
10. Bedroom
11. Bathroom

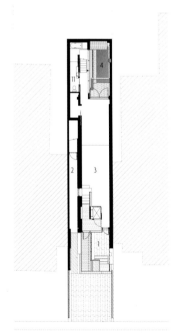

Studio

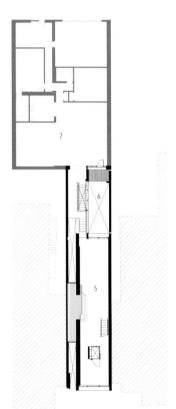

Gallery

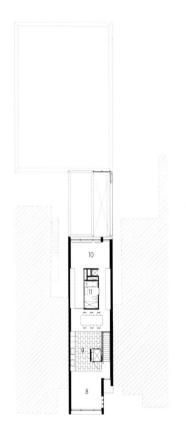

Residence

Vertical House

This 2,400 square-foot residence on Pacific Avenue in Venice, California, cleverly responds to the site restrictions: extremely narrow setbacks and selected views. A simple material—cement fiber board—has been innovatively used in conjunction with three types of colored glazing, thus defining architecture through the envelope of a volume rather than through the volume itself.

Architect: Lorcan O'Herlihy Architects
Location: Venice, CA, USA
Date of construction: June 2004
Photography: Undine Pröhl

Sketch

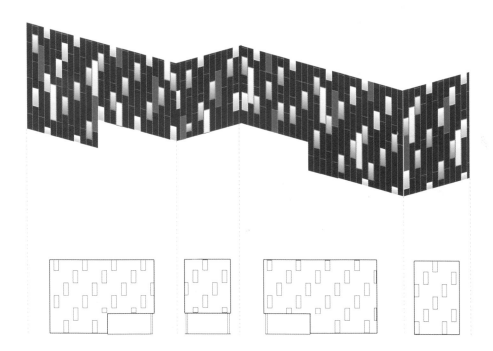

Exterior envelope diagram

Site plan

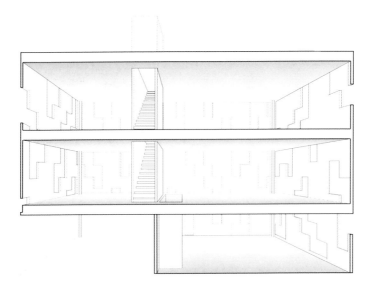

Floor plate diagram

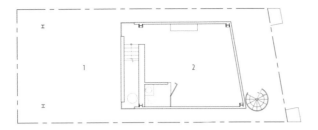

First floor

1. Carport
2. Studio
3. Master bedroom
4. Guest bedroom
5. Bathroom
6. Closet
7. Dining room
8. Kitchen
9. Living room
10. Cabinet
11. Laundry room
12. Glass pavilion
13. Roof terrace

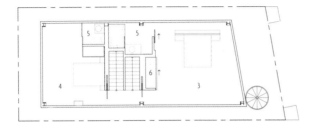

Second floor

Third floor

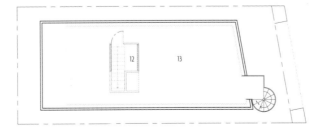

Roof

Circulation and program are defined through a centrally located vertical core. This core contains a central stair which extends to the roof.

Three types of colored glazing are used in the structure's façade, with an interesting effect on the interiors.

Threefold House

Architect: Takao Shiotsuka
Location: Kumamoto, Japan
Date of construction: 2007
Photography: Toshiyuki Yano/
Nacasa & Partners

Located in a high density residential area, this house for a family of six could only receive sunlight from southeast-facing spaces. In order to obtain as much sunlight as possible, Takao Shiotsuka designed a small building on three layers. A southeast-facing wall retreats as it turns into the upper floor, thus allowing each floor to receive as much sunlight as possible.

North elevation

West elevation

South elevation

East elevation

In order to receive sunlight comfortably throughout the house, the structure is stacked on three levels, retreating slightly as it reaches the upper floor.

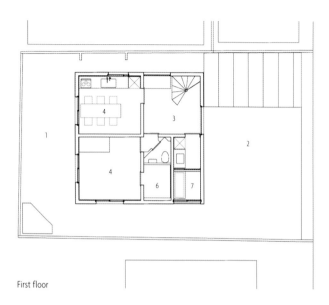

First floor

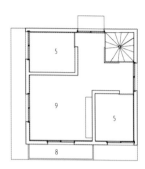

Second floor

Third floor

Roof

1. Garden
2. Parking area
3. Entrance hall
4. Kitchen/dining room
5. Bedroom
6. Closet
7. Bathroom
8. Terrace
9. Living room
10. Study

Architect: Mikan
Location: Tokyo, Japan
Date of construction: 2007
Photography: Covi

Facing a green pedestrian path and locked between houses and apartment buildings, the general volume of this house was notched and tilted according to legal setback lines and vertical courtyards. A giant veil of expanded metal allows the inner space to be expanded and provides privacy at the same time. Developed over three floors, each level has a distinctive characteristic.

Transversal section

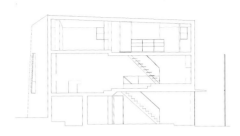

Longitudinal section

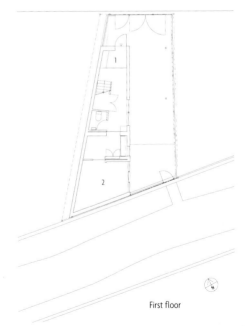

First floor

Second floor

Third floor

1. Entrance
2. Bathroom
3. Living room/dining room
4. Bedroom 1
5. Bedroom 2

Each floor has a distinguishing feature: the ground floor has concrete, cement and white walls; the second floor dark; the top floor white and clear again.

The L shape of the floor plan
and the skipping floors create
a gentle flowing space. The
family space is clad in
wooden panels tinted dark to
create a cozy atmosphere.

Reflection of Mineral

The unusual shape of this house is the result of a limited building site, architectural regulations and the client's wish for a fun, maximum-sized house with a roofed parking space. Cut in the shape of a diamond, this house stands out from the others in this dense residential neighborhood. The white finish gives it an abstract quality while the angular surfaces reflect sunlight and shadow, constantly transforming the house's appearance.

Architect: Atelier Tekuto
Location: Nakano-ku, Tokyo, Japan
Date of construction: October 2006
Photography: Makoto Yoshida

Site plan

This small diamond-shaped
house in a dense Tokyo
residential neighborhood
provides maximum interior
space and light.

Erected on a limited building
site (485 square-foot), this
house fulfills the client's
wishes, including a roofed
parking space.

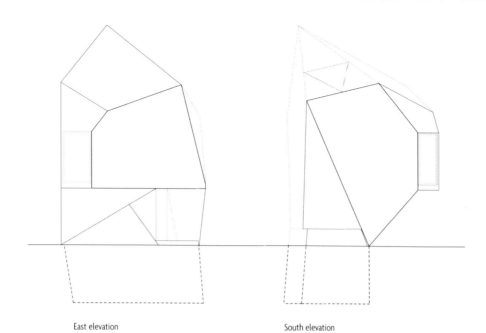

East elevation

South elevation

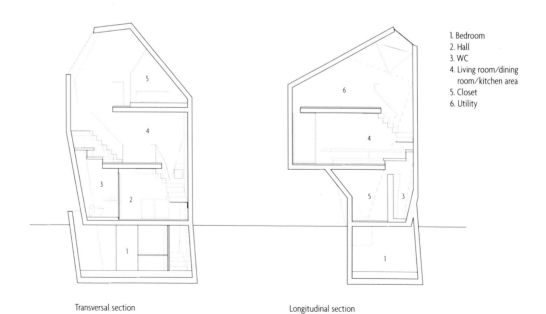

1. Bedroom
2. Hall
3. WC
4. Living room/dining room/kitchen area
5. Closet
6. Utility

Transversal section

Longitudinal section

This house is a polyhedron which holds a variety of surfaces which create a kind of optical illusion inside.

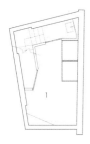

Basement

First floor

Second floor

Third floor

1. Bedroom
2. Hall
3. Closet
4. WC
5. Porch
6. Parking
7. Living room/dining room/kitchen area
8. Utility room
9. Bathroom
10. Top light

Alonso-Marmelstein House

Architect: Alonso Balaguer
& Arquitectos Asociados
Location: Barcelona, Spain
Date of construction: 2005
Photography: Jose María
Molinos, Pedro Pegenaute,
Pere Peris

Finding a site in the city is not an easy task nowadays and more often than not entails complex construction characteristics, which call for complicated layouts. This 30-foot plot houses an original home spread out over six floors, formed by a succession of interior patios which helps maintain a visual continuity of this vertical space.

Roof

Third floor

Second floor

Mezzanine

First floor

Basement -1

Basement -2

Roof

Third floor

Second floor

Mezzanine

First floor

Basement -1

Musitu Street 49

Bertrán Street 80

Longitudinal section 1

Roof

Third floor

Second floor

Mezzanine

First floor

Basement -1

Roof

Third floor

Second floor

Mezzanine

First floor

Basement -1

Basement -2

Bertrán Street 80

Musitu Street 49

Longitudinal section 2

Bertràn Street 80   Musitu Street 49

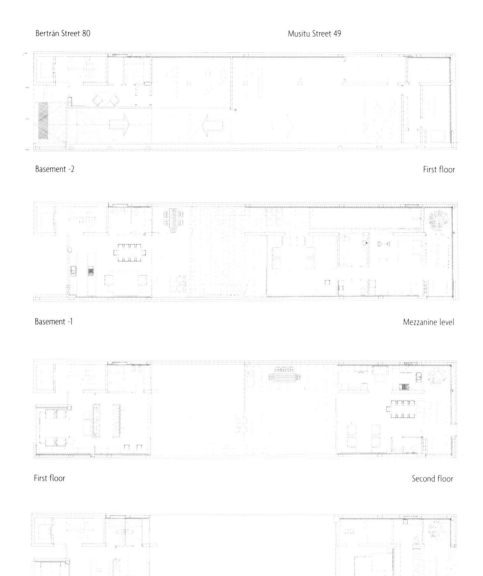

Basement -2   First floor

Basement -1   Mezzanine level

First floor   Second floor

Second floor   Mezzanine

Bertrán Street 80

Musitu Street 49

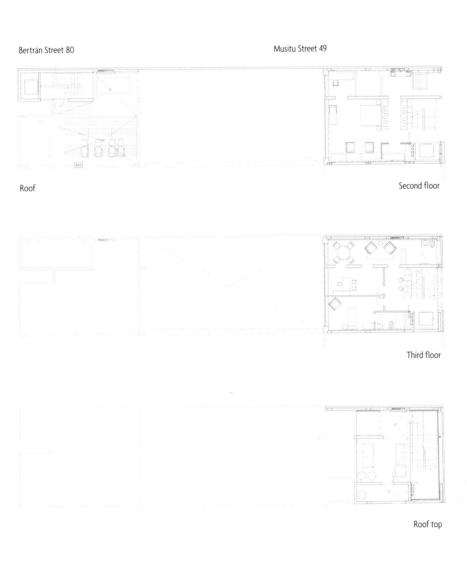

Roof

Second floor

Third floor

Roof top

One of the most important
features of this house is its
transparency. Communication
and relation between the
spaces is possible thanks
to translucent surfaces and
transparent paving.

To make the most of available natural light, the basement has a skylight, so that all levels of the house can be seen and receive plenty of light, vertically.

A former graphic arts workshop in the Gracia neighborhood of Barcelona was turned into an urban single-family home, with a photographic studio in the basement. The building consists of a basement, ground floor, mezzanine, first floor and attic, with a car lift to avoid the kind of surface reduction that a vehicle ramp would cause.

House in Barcelona

Architect: Luis Felipe Infiesta Calzado/SSCV Arquitectos
Location: Barcelona, Spain
Date of construction: 2006
Photography: Montse Campins, Luis Felipe Infiesta

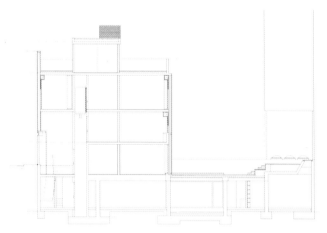

Longitudinal section

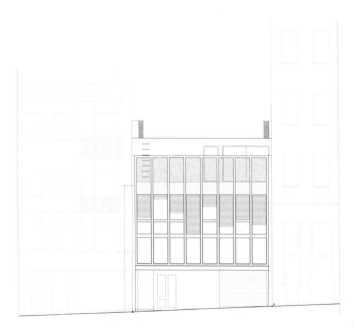

Façade

The façade is exclusively
made of steel and wood, with
shutters which filter the light
in a discreet way. Graffiti by
Lucas Milá marks the
pedestrian entrance.

Basement

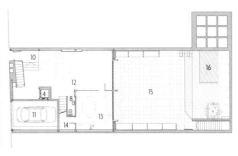

First floor

First floor (Mezzanine)

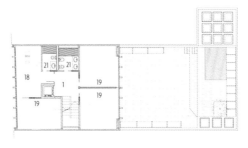

Second floor

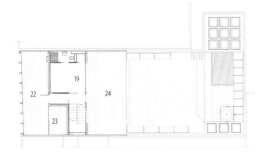

Attic

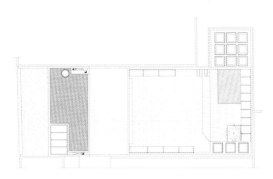

Roof

1. Lobby
2. Machine room and elevator platform
3. Laundry room
4. Elevator
5. Garage
6. TV room
7. Laboratory
8. WC
9. Photographic studio
10. Hall
11. Platform
12. Living room
13. Kitchen/dining room
14. Larder
15. Patio
16. Pond
17. Storage room
18. Study
19. Bedroom
20. Dressing room
21. Bathroom
22. Clothes line
23. Boiler room
24. Terrace

A mezzanine level between
the ground floor and the first
floor contains the master
bedroom, dressing room and
study/library.

Landscape & Topography

New House in Brione

Built in a dense neighborhood, this new house offers views over the lake and surrounding mountains. The project solves the problem of creating a home in urban chaos. Two simple stone boxes placed against one another and protruding from the mountain give this structure a classic quality and make it more part of the landscape than of the neighborhood.

Architect: Markus Wespi Jérôme
de Meuron Architetti
Location: Brione sopra Minusio,
TI, Switzerland
Date of construction: 2005
Photography: Hannes Henz

Despite the dense
neighborhood, this house has
come out a winner. Built into
the mountain, the result is a
classic structure which has
become part of the landscape.

Site plan

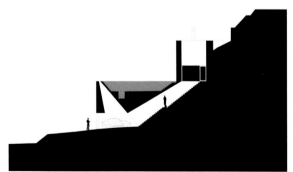

Transversal section

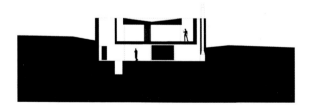

Longitudinal section

1. Entrance
2. Parking
3. Stairs
4. Mechanical room
5. Void above parking
6. Courtyard
7. Entrance
8. Closet
9. Kitchen
10. Dining area
11. Living room with chimney
12. Cupboards
13. Lumber room
14. WC
15. Storage room
16. Garden
17. Swimming pool
18. Bedroom
19. Shower
20. Shower/WC
21. Courtyard
22. Void above living room
23. Void above courtyard

First floor

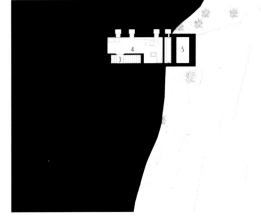

Second floor

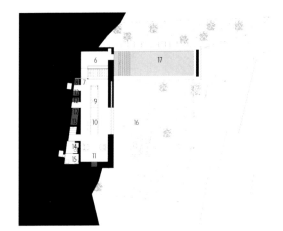

Third floor

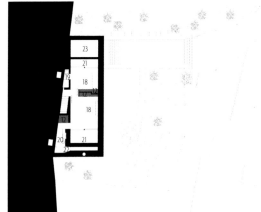

Fourth floor

Extra light comes in through
an interior courtyard and
illuminates the private areas.

Light colors and simple,
straight lines create a
soothing interior.

White Cave

Architect: Takao Shiotsuka
Location: Oita, Japan
Date of construction: February 2007
Photography: Toshiyuki Yano/ Nacasa & Partners

This house is built on an irregularly-shaped plot with neighboring land to the north, thick trees on the west and south sides, and a view to the town below on the east. Built on a hill, the design of the house incorporates the six-and-half-foot drop. Consequently, the scenery inside changes similarly to the one experienced from the outside.

The building adapts to the
six-and-a-half-foot drop of this
peculiar site, both on
the outside as well as
on the inside.

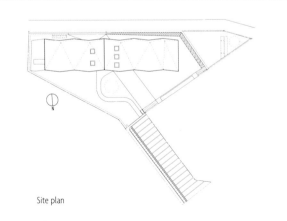

Site plan

The walls and windows shift in their orientation, creating an interesting pattern and relation between the interiors and the surrounding landscape.

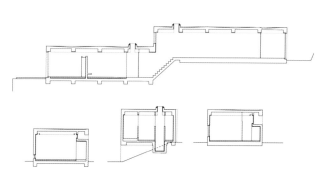

Longitudinal and transversal sections

The all white interiors are in stark contrast with the rough concrete structure and the view of the sprawling city below.

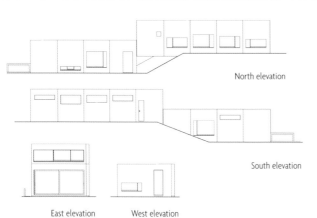

North elevation

South elevation

East elevation   West elevation

To counterbalance the angles created within as a result of the exterior structure, the interior design is kept to a minimum using predominantly white and smooth, straight lines.

Floor plan

1. Entrance
2. Lounge
3. Kitchen
4. Dining room
5. Bathroom
6. Bedroom

Concrete House

Architect: María Victoria
Besonías, Guillermo de Almeida,
Luciano Kruk/
BAK Arquitectos
Location: Mar Azul, Buenos Aires,
Argentina
Date of construction: 2006/2007
Photography: Daniela Mac Adden

This summer house is situated in a lush forest of conifers 400 km south of Buenos Aires. Concrete was the obvious material, which would solve the few limitations of this project: a limited budget, low impact on the surrounding landscape, as little subsequent maintenance as possible and a short construction period.

The southwestern façade,
buried in the sand, has an
aperture from one extreme to
the other, while the
northwestern façade enjoys
floor-to-ceiling windows so
the views ahead can be
enjoyed.

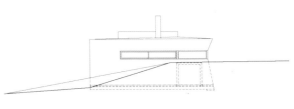

Northwest elevation

Transversal section

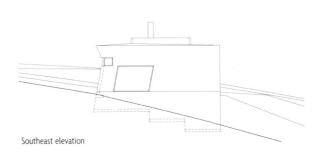

Southeast elevation

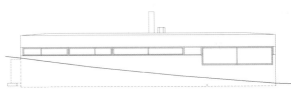

Southwest elevation

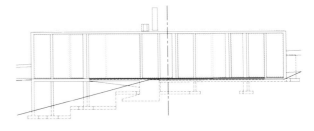

Longitudinal section

Northeast elevation

The domestic areas are protected from unwanted prying eyes by walls placed at varying intervals according to the need of the area.

1. Living room
2. Dining room
3. Kitchen
4. Fireplace
5. Master bedroom
6. Master bathroom
7. Bedroom
8. Bathroom
9. Deck

Floor plan

The bespoke furniture is made
of recycled Canadian pine
wood (formerly crates for
packaging engines) and the
same concrete used for
the structure.

There is no main entrance to
this house. The house can be
accessed through any one of
the living areas.

Gradman House

This small country residence is located on a steep up-sloping lot surrounded by mature oak, fir and bay trees. The owners wanted to promote the enjoyment of the outdoors, maximize views and sensitively knot the house to the land. Designed on five floor levels, each level gently steps with the topography and creates a distinct zone for living.

Architect: Swatt Architects
Location: Inverness Park, CA, USA
Date of construction: 2005
Photography: César Rubio

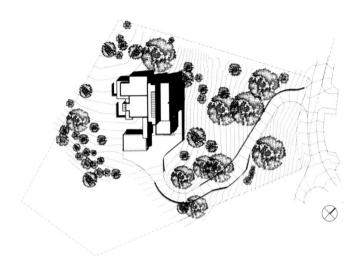

Site plan

Floor plan

1. Entrance
2. Kitchen
3. Dining room
4. Living room
5. Laundry room
6. Bedroom
7. Bathroom
8. Master bedroom
9. Master bathroom
10. Closet
11. WC
12. Garage

The entry and circulation space is designed as a light-infused central spine, which joins the lower public spaces with the private areas above.

The living and dining areas open to expansive terraces with views over the bay and wetlands below.

The house is designed in such
a way that each level follows
the topography of the site,
thus creating five levels and
distinct domestic zones.

Each private space opens out to its own
private hillside garden and the top of the
site.

Goulet House

Architect: Saia Barbarese
Topouzanov Architectes
Location: Sainte-Marguerite-du-
lac-Masson, Canada
Date of construction: 2003
Photography: Marc Cramer

A steep slope descends from north to south towards the lake and is interrupted by a sheer rock face shielding a plateau. Sited on this plateau in a verdant forest between erratic stones is Goulet House, a long house isolated in a harsh landscape, which has adopted the form dictated by the projecting ledge.

29

Fir plywood skin lines the interior of this house creating a warm cozy atmosphere. An intimate family zone occupies the full height of the building.

Elevations

1. Living room
2. Terrace
3. Office
4. Storage room
5. Kitchen
6. Bedroom
7. Bathroom
8. Master bedroom
9. Mezzanine
10. Corridor
11. Vestibule

Transversal sections

Longitudinal sections

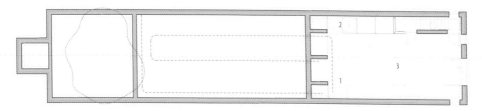

First level

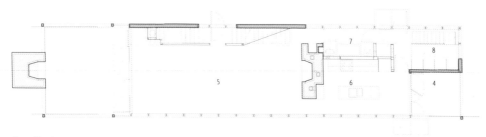

Second level

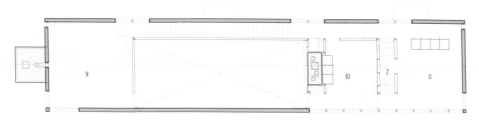

Third level

1. Atelier
2. Storage
3. Shower
4. Terrace
5. Living room
6. Kitchen
7. Bathroom
8. Vestibule
9. Office
10. Bedroom
11. Master bedroom

The design of this house situated on the west coast of Sweden is basically three horizontal volumes connected to each other, two of them stretching out on the ground and a smaller one looking over a height to view the seat. The architect chose different volumes and materials, patterns and colors to make the house seem smaller.

Architect: Hulting Arkitekter
Location: Gothenburg, Sweden
Date of construction: 2007
Photography: Ulf Celander

1. Entry
2. Living room
3. Dining
4. Kitchen
5. Cool storage
6. Store
7. Bathroom
8. Bedroom
9. Play, music and
   guest room
10. Laundry and technique
11. Toilet
12. Fireplace
13. Wood store for
    fireplace and sauna
14. "Hole in the wall"
    to store small things
15. Stone terrace for
    external dining
16. Entrance yard
17. Green house
18. Dressing
19. Sauna
20. Hot tub, shower, relax
21. Stone wall
22. Green roof
23. Library, TV, study
24. Balcony
25. Roof terrace
26. Wooden path
27. Sitting bench

First floor

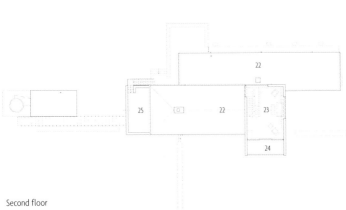

Second floor

30

Seen from a distance, this single villa looks like a group of houses, because of the use of different volumes, materials, patterns and colors.

The large combined living
room and kitchen
(968 square-foot) is
the heart of the house.
Though the space is large,
the architect has created an
intimacy in the room.

The interiors are kept simple
with light colors and windows
and doors made of pine wood
painted in white.

The single room on the
second level is a calm room
for reading, work and TV,
focused at the two views of
the sea to the southwest and
north to northeast.

Delta Shelter

This 1,000 square-foot weekend cabin in Washington State is situated near a river in a floodplain in a wooded valley. Made of steel, this home is built on stilts and the domestic program is organized over three levels with the living room and kitchen at the top. Large 10-by-18-foot steel shutters can be opened or closed by hand crank.

Architect: Tom Kundig/Olson Sundberg Kundig Allen Architects
Location: Mazama, WA, USA
Date of construction: 2005
Photography: Undine Pröhl

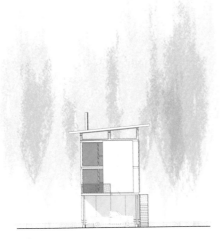

East elevation

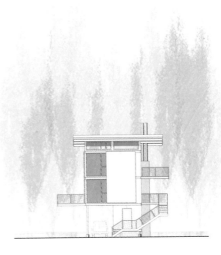

North elevation

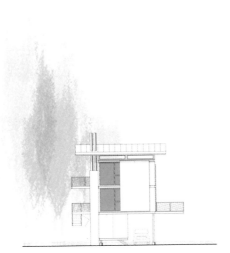

South elevation

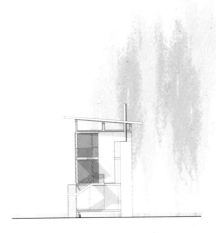

West elevation

The living room and kitchen are at the top of this steel cabin, which can be opened or closed by using a hand crank to move the steel shutters.

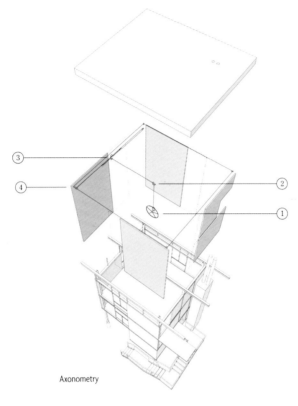

Axonometry

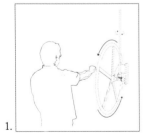

1.

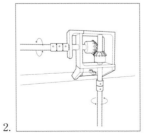

2.

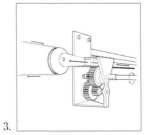

3.

4.

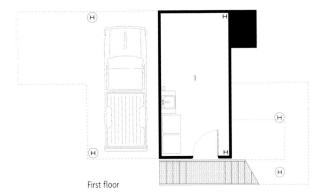

First floor

1. Storage
2. Master bedroom
3. Guest room
4. Bathroom
5. Powder room
6. Dining room
7. Living room
8. Kitchen

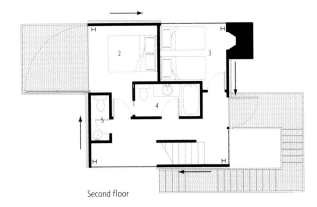

Second floor

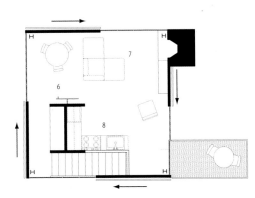

Third floor

This beach house is situated on a site which abuts the foreshore and faces east and the sea. The design was developed in response to the site constraints and limited budget and the client's desire for a simple, strong and comfortable family beach house, open in spirit yet also private. The result is an enigmatic box pulled with the cleft providing protected open space, access and egress, sunlight, views and circulation.

Eastern Beach House

Architect: Clinton Murray,
Shelley Penn/Clinton Murray
Architects
Location: Port Fairy, Victoria,
Australia
Date of construction: 2007
Photography: Trevor Mein

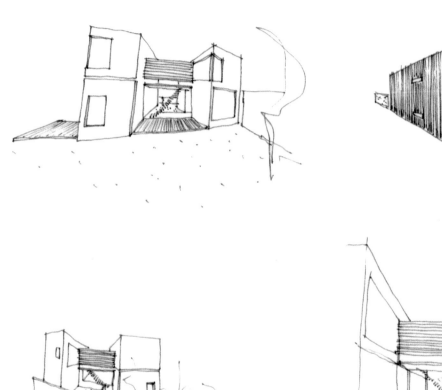
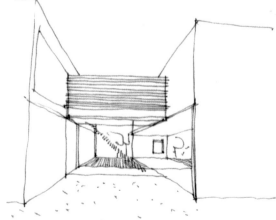

Sketches

The simplicity of this floor plan results in a house where every room has access to direct sunlight and views of the beach with varying degrees of privacy and protection.

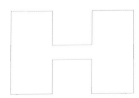

Site plan

Vertical timber cladding
characterizes the external,
closed shell of the "box,"
thus presenting a more
imposing and hermetic form
to the outer edges.

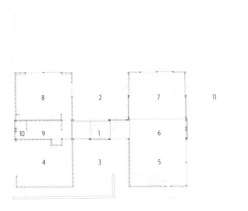

8	2	7	11
10 9	1	6	
4	3	5	

First floor

12		12
14	15	13
12		12

Second floor

1. Entry
2. Deck 1
3. Courtyard
4. Garage
5. Living room
6. Dining room
7. Kitchen
8. Study
9. Laundry room
10. WC
11. Deck 2
12. Bedroom
13. Ensuite bathroom
14. Bathroom
15. Link

This house is a retreat with layers of transparency, enclosure and visual/physical connection, with the sea always as a reference.

Villa Tjuvkil

This lot right by the sea could not be reached by road so all construction material had to be shipped. Due to these difficulties, the designed called for a simple, wooden house erected on the bedrock without any blasting. The principle of simplicity and low maintenance was applied to the design of this house built on a limited 861 square-foot.

Architect: Karin Windgårdh/
Windgårdh Arkitektkontor
Location: Tjuvkil, Sweden
Date of construction: 2003
Photography: Ulf Celander

East façade

North façade

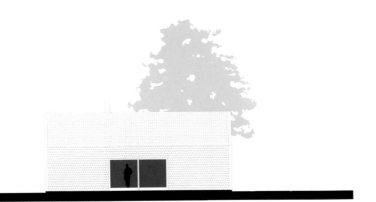

South façade

West façade

The situation right by the sea
and the open view towards the
sunset directed the principles
of this small vacation home on
the Swedish coast.

1. Entrance
2. Kitchen
3. Dining room
4. Living room
5. Bedroom
6. Toilet
7. Void
8. Bedroom

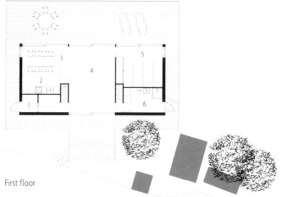

First floor

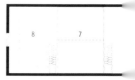

Second floor

The walls and roof of this
house are covered by
shingles made of Canadian
cedar, which is left untreated
and will adapt to the colors of
the rocks.

The open interior is supported by a steel structure. A large boardwalk expands the living room and large floor-to-ceiling windows blend the interior with the exterior.

Villa Berkel

Architect: Architectenbureau
Paul de Ruiter
Location: Veenendaal,
The Netherlands
Date of construction: 2005
Photography: Pieter Kers

Surrounded by dark woods, it was important to ensure as much light as possible could enter the villa without forgoing privacy. Therefore the building plot is divided into three long strips at right angles to the road. The bottom and southernmost strip is reserved for the garden, the middle strip contains the villa and the most northerly strip provides access to the house (entrance, parking, drive). With this layout residents can keep certain parts of the house private and out of sight.

The house is entirely focused
on the secluded garden to the
south, thus allowing
the residents to feel they are
out of doors at all times.

Site plan

Three of the four façades are made of glass and every room in the villa looks directly out onto the garden.

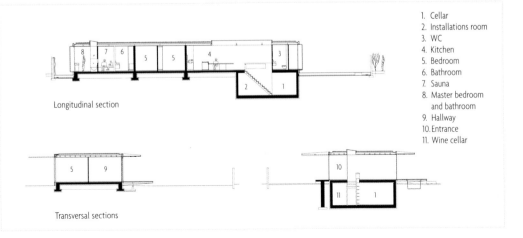

Longitudinal section

Transversal sections

1. Cellar
2. Installations room
3. WC
4. Kitchen
5. Bedroom
6. Bathroom
7. Sauna
8. Master bedroom and bathroom
9. Hallway
10. Entrance
11. Wine cellar

Each function has its own
zone in the house and can be
cut off by means of
translucent sliding walls. The
character of the functions
gradually becomes more
intimate.

The house is divided into three strips: the eastern sections contains the more public functions, while the western section is reserved for more intimate activities.

1. Entrance
2. Kitchen
3. Living room
4. Study
5. Bedroom
6. Master bedroom
7. Master bathroom
8. Pond
9. Terrace

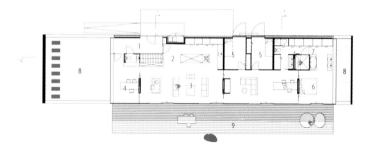

Floor plan

Sustainability

This house is designed in such a way to achieve a higher sense of spatial freedom within an architectural home. Divided into public below and private above, the house also has a parent's side and a children's side, allowing each user their privacy and sound control. The house also has a solar voltaic on the roof and a gray water reclamation drip irrigation system, among other sustainable aspects.

Nelly Residence

Architect: Trevor Abramson, Douglas Teiger/Abramson Teiger Architects
Location: Los Angeles, CA, USA
Date of construction: 2006
Photography: Richard Barnes

Trespa wood panels offset the
front wall with an air barrier,
providing cool air circulation,
as does a reflecting pond at
the face of the glass wall.

1. Entry
2. Living room
3. Pantry
4. Powder room
5. Kitchen
6. Garage
7. Family room
8. Breakfast room
9. Guest bedroom
10. Guest bathroom

First floor

Second floor

11. Master bedroom
12. Master bathroom
13. Master closet
14. Roof
15. Bedroom
16. Closet
17. Bathroom
18. Laundry room
19. Mechanical room
20. Playroom
21. Storage

The house is divided into four boxes. Each box is individually articulated and expressed by the use of a different exterior surface treatment.

Alisal Residence

Though not immediately visible, this house is an environmentally sustainable house, in terms of its relationship to the site, reuse of the foundation, built-in devices, photovoltaic heat capture, radiant floor heating and natural and sustainable materials. Moreover, it is packed into 3,200 square-foot, yet has all the elements of a 5,000 to 6,000 square-foot dwelling.

Architect: Shubin & Donaldson
Architects
Location: Santa Barbara,
CA, USA
Date of construction: 2006
Photography: Ciro Cohelo

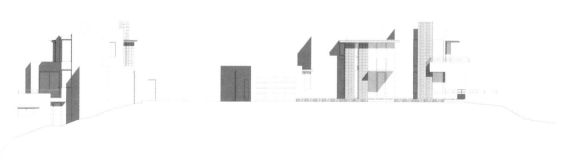

North elevation

West elevation

South elevation

East elevation

Thanks to a neutral color scheme, the colors of nature can become the predominant palette.

The house is designed in a way so that it is constantly open to the outdoors. An infinity pool just outside the living room furthers the idea of expanse.

1. Entrance
2. Living room
3. Dining room
4. Kitchen
5. Outdoor dining room
6. Powder room
7. Outdoor lounge
8. Pool
9. Garage
10. Master bedroom
11. Master bathroom
12. Balcony
13. Guest bedroom
14. Guest bathroom
15. Gym

First floor

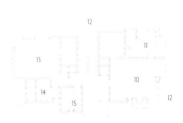

Second floor

This house takes up minimal space as well as minimal resources. Whatever could be reused was incorporated into the building.

This 204.5 square-foot dwelling is a prototype building, a proposal for more economical and ecological housing. With four rooms covering the basic living functions, the project focuses on quality of space, material and natural light and tries to reduce unnecessary floor area. The result is a dwelling a quarter of the price of any same-size apartment.

Architect: Sami Rintala
Location: Galleri ROM, Oslo, Norway
Date of construction: August 2007
Photography: Sami Rintala

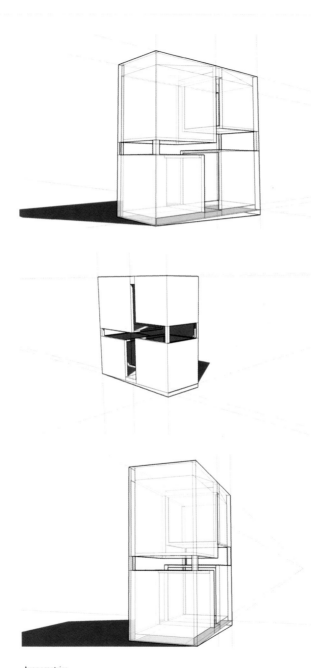

Axonometries

Boxhome is a peaceful urban home which covers the basic living functions in four rooms and 204.5 square-foot.

House Wesolowski

The architect tried to respect the traditional structure of the regional farm houses of French-speaking Bavaria. The final design is a reinterpretation of the traditional style, including a slanted roof, minimal details on the façade and a predominance of natural materials. Energetically-speaking, the home requires very little energy for climatization.

Architect: Att Architekten
Location: Gräfenberg, Germany
Date of construction: 2006
Photography: Stefan Meyer

The design of this house is based on the
traditional rural style of Bavaria, Germany.

The house enjoys natural
ventilation in the summer and
relies on a renewable source
of energy to heat the rooms
in winter.

The different domestic areas have been
organized around a central patio, where
an existing cherry tree was preserved.

Rainwater is collected in a tank, purified
and used for the washing machine,
bathroom and garden.

Mountain Retreat

Architect: Resolution:
4 Architecture
Location: Kerhonkson,
NY, USA
Date of construction: 2005
Photography: Floto & Warner

With its angular lines, soaring height and unique blend of warm cedar siding with cool gray concrete panels and glass, this home is carefully crafted into its unique surroundings. The house has a butterfly roof which channels rainwater to two stainless-steel scuppers and sustainable engineered bamboo floors were employed throughout.

The concrete stilts on which this house
partially sits act as support for the great
room above and define the parking
spaces for an uncluttered entry and
carport below.

Site plan

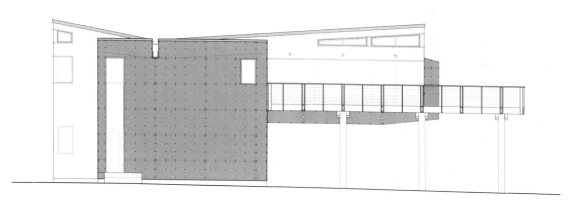

North elevation

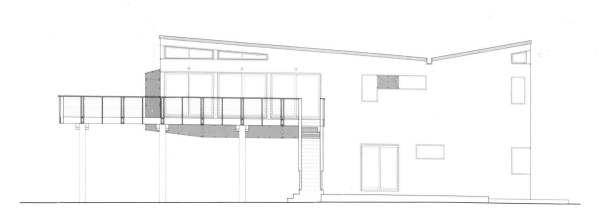

South elevation

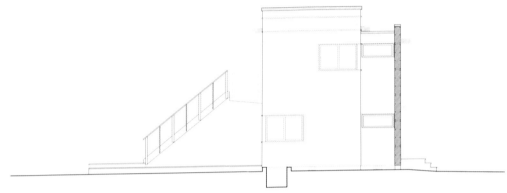

East elevation

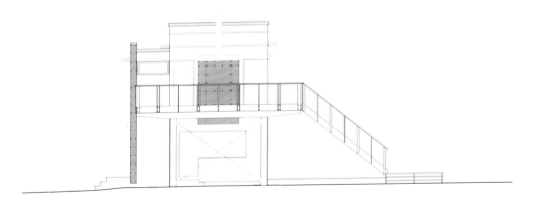

West elevation

50

An elevated cedar deck wraps around three sides of the main living area, thus allowing the entire room to be opened to the outdoors with ease.

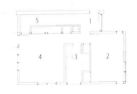

1. Entry
2. Bedroom
3. Bathroom
4. Flex space
5. Mechanical/storage room
6. Covered carport
7. Rear deck

First floor

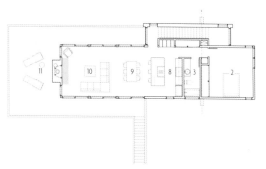

8. Kitchen
9. Dining room
10. Living room
11. Upper deck

Second floor

Sustainable bamboo floors (preserved
with whitewash) add a durable yet
softening touch to predominantly
light-hued interiors.

Kropach Catlow Farmhouse

This stripped composition open to, yet protected from, the elements is based on an east-west axis to take in the views of Mount Warning. Made of a steel frame and softly burnished steel cladding, glass louvers are used extensively to generate cross-ventilation. An extended ranking roofline to the north provides a shield from the summer sun and captures winter light.

Architect: Bligh Voller Nield
Location: Yocum, NSW, Australia
Date of construction: 2003
Photography: Peter Hyatt

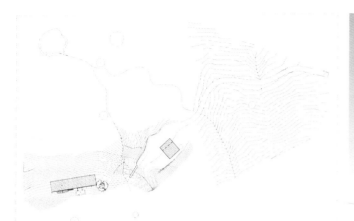

Site plan

This stretched veranda is a model of ergonomic and democratic design; a simple diagram and plan reinforced by a steel frame and softly burnished steel cladding.

Like a glass box, the interior and exterior
can be blurred with sliding shutters. To
the north, an extended roofline captures
and shields the sunlight.

1. Living room
2. Dining room
3. Kitchen
4. Master bedroom
5. Master bathroom
6. Bathroom
7. Bedroom

Floor plan

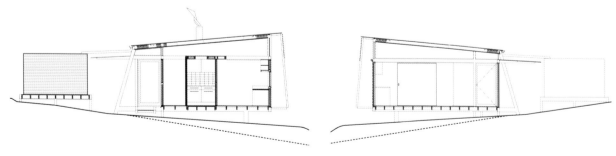

Transversal sections

A deck on the western end provides highly useable additional floor space. Adjustable louvers define the level of breezeway comfort control.

Based on an east-west axis to take in the views of the surrounding landscape, this extruded form allows thermal efficiencies and construction economies.

Sliding concertina walls/
windows blur the inside and
outside experience, while
adjustable broad-bladed
louvers help control cross-
ventilation.

An elevated platform incorporates two
bedrooms and provides greater privacy
with views predominantly to the north.

Modewarre House

This house challenges the notion of what a typical, single level "colonial" house usually is: located on large plots, linear in plan and oriented along an east-west axis to maximize solar gain. This house is orientated along a north-west-to-east axis, has a solid structure to the south and is open to the north.

Architect: Saaj Design
Location: Modewarre, Victoria,
Australia
Date of construction: 2006
Photography: Patrick Redmond

Situated on the crest of a small mound within an open plane of a large stud farm, this house is perceived as a windbreak.

Site plan

The orientation of the house
has been designed to
maximize solar access and
windows have been
strategically placed to
encourage cross-ventilation.

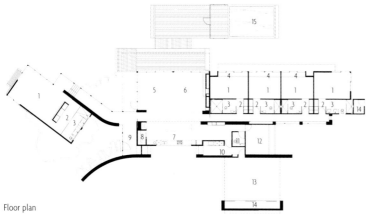

Floor plan

1. Bedroom
2. Walk-in-closet
3. Bathroom
4. Daybed
5. Dining room
6. Living room
7. Kitchen
8. Pantry
9. Entry
10. Laundry room
11. Powder room
12. Study
13. Carport
14. External store
15. Pool

This house is situated in a verdant Brisbane gully subject to severe flooding in periods of intense rainfall. As a result, the living areas are raised above flood level. Because of the fragile nature of the soil, a lightweight construction system is supported on a grid of steel columns with a minimal number of footings. Thus, the overland flow can continue under the house and the disturbance to the existing tree root systems is as minimal as possible.

Gully House

Architect: Shane Thompson/Bligh Voller Nield; Danny Fox/Daniel R. Fox Architect
Location: Brisbane, Queensland, Australia
Date of construction: 2004
Photography: David Sandison

There was no significant bulk excavation for the construction of this house. The house was carefully placed between and around existing significant trees.

A canopy of trees not only provides shade, it is also an ideal way of creating privacy from the close proximity of neighbors to the west.

The house is situated in a steep gully and bridges over an existing overland flow, but is designed in a way to allow the existing overland flow to continue under the house toward the river.

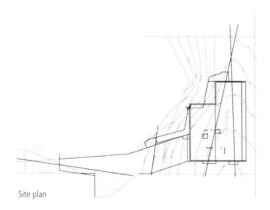

Site plan

The external walls of the house are fully insulated and clad with plantation-grown "eco-ply," and eave-overhangs and weather screens protect all external windows and doors from the elements.

1. Storage
2. Cars
3. Rainwater tank
4. Lower entry
5. Driveway

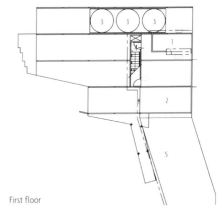

First floor

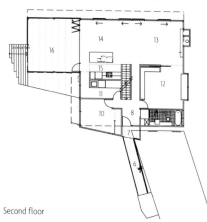

Second floor

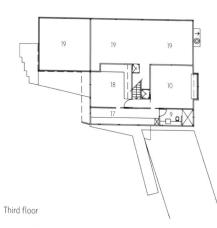

Third floor

6. Walkway entry bridge
7. Upper external entry
8. Upper internal entry
9. Bathroom
10. Bedroom
11. Laundry room
12. Library
13. Living room
14. Dining room
15. Kitchen
16. Verandah
17. Walk-in robe
18. Studio
19. Void

Plantation-grown "blackbutt" timber flooring is used throughout the interior and, as far as possible, materials have been chosen for their ability to be recycled in the future.

60

Large openings between wall spaces facilitate a high level of cross-ventilation along with thermal ventilation afforded by high level roof openings.

Essex Street House

Architect: Andrew Maynard
Location: Brunswick,
Melbourne, Victoria, Australia
Date of construction: 2006
Photography: Peter Bennetts,
Dan Mahon

The brief for this residential alteration and extension to an existing double-fronted weatherboard house required two bathrooms, a bedroom, living area, kitchen and an increased connection with the outside areas. The response to the brief was that any addition should run along a southern boundary to maximize solar access to the new and existing spaces.

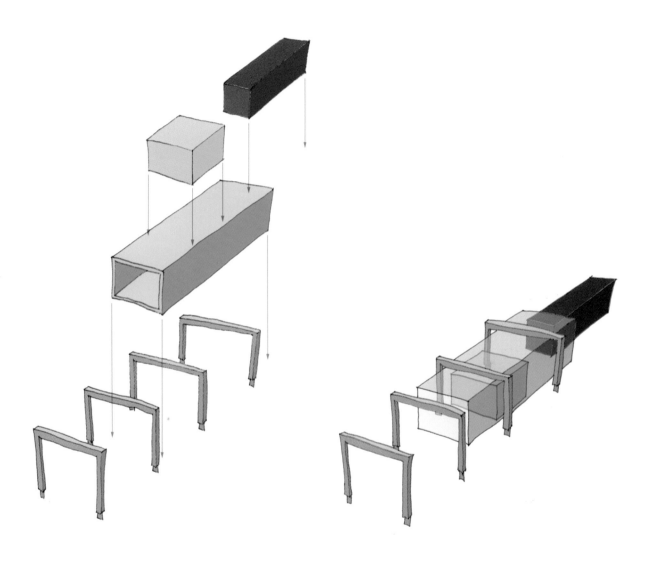

Concept diagram

61

Insulation, a sheltering context and well-designed sunshading makes the design an efficient home without the use of elaborate tactics or expensive equipment.

Cape Schank House

Architect: Paul Morgan
Architects
Location: Cape Schank, Victoria,
Australia
Date of construction: 2006
Photography: Peter Bennetts

This house is located in an area near rugged coastline subject to strong prevailing winds and sits within an expanse of native tea trees. The shell of the house was developed as a result of the analysis of sunlight movement and wind frequency, speed and direction, and the modeling of the wind onto the site.

The "underskin" flows
continuously from the
external eaves to the ceiling
and is "gathered" into a bulb
tank.

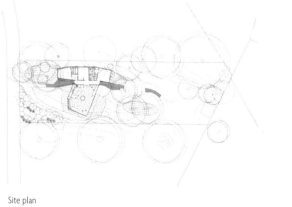

Site plan

Wind scoops on the south
elevation trap cooling winds
during summer while
providing shade from the
hot afternoon sun.

A bulb tank cools the ambient air temperature of the living room during summer, supplies rain water and structurally carries the roof load.

1. Living room/dining room/ kitchen
2. Terrace
3. Bedroom
4. Bathroom

Floor plan

Lovely Ladies Weekender

Three long-term friends came up with an alternative plan which would give them the lifestyle they wanted at a fraction of the cost. The brief was to create an inexpensive home responsive to sustainability issues and sparing as much remnant bushland as possible. The result is a narrow, two-storey rectangular house with bedrooms and bathrooms downstairs and lounge, dining and kitchen upstairs.

Architect: Marc Dixon
Location: Walkerville, Victoria, Australia
Date of construction: 2006
Photography: Lucas Dawson

To keep construction costs down and reduce disturbance to the surrounding native fauna, the size of the footprint is small.

East elevation

North elevation

The narrow floor plan enhances the opportunity for cross-ventilation and maximizes northern orientation, important for passive solar access in the Southern Hemisphere.

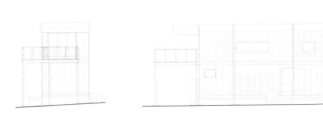

West elevation

South elevation

Water tanks harvest the
rainwater for reuse on site.
Prefinished Colorbond® steel
sheet is used for the building
and water storage tanks.

First floor

Second floor

1. Entrance
2. Deck
3. Deck area over
4. Tank
5. Stairs
6. Storage
7. Bedroom
8. Bathroom
9. Shower room
10. Stairs
11. Open deck
12. Dining room
13. Kitchen
14. Living room

68

A stand of trees provides a level of privacy to the lower deck accessed from the bedrooms.

Extensions, Remodels, Renovations

This light frame wood construction sits on top of an existing single-family house. The north side of the rooftop was opened up to create a panoramic windowfront. Floor-to-ceiling windows not only offer views over the garden below but also catch a maximum amount of sunlight. The street side is filtered by wooden louvers and appears closed at first.

Architect: Share Architects
Location: St. Pölten, Austria
Date of construction:
2004/2005
Photography: Franz Ebner

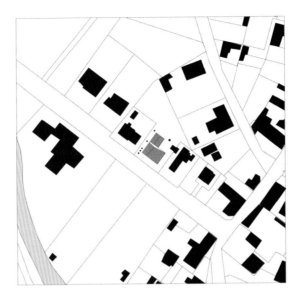

Site plan

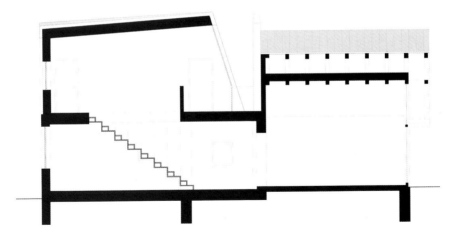

Longitudinal section

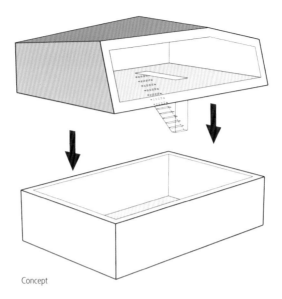

Concept

1. Entrance
2. Living room
3. Kitchen
4. Bathroom
5. Guest room
6. Master bedroom
7. Master bathroom
8. Children's room

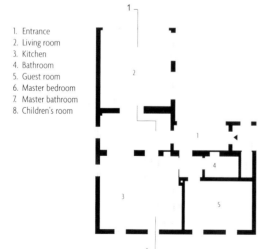

First floor

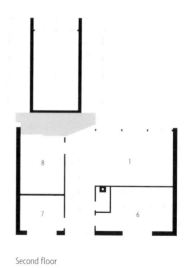

Second floor

69

A hanging staircase connects the upper level with the ground floor and functions as a separation between the kitchen and dining areas in the existing house.

Wooden louvers at slightly different distances allow plenty of privacy for the inhabitants without the latter feeling imprisoned.

**Trichterweiterung/
Funnel Extension**

The *trichter* is an extension to an existing 1930s house. The old house was converted into two different apartments, one on the ground level and another on the first floor. The funnel extension is attached to the owner's ground floor apartment creating one big open space connecting the old with the new via the kitchen.

Architect: Peanutz Architekten
Location: Berlin-Dahlem,
Germany
Date of construction: 2005
Photography: Stefan Meyer

The windows of the new
studio, where the owner
paints, consist of two slits
exposing light underneath the
ceiling and on the floor.

The artist's ground floor
apartment has been
restructured through large
openings in the walls to
create one open space
containing the kitchen and a
studio.

There are no openings in the wall facing the street, in order to enhance the difference between the new object and the house.

Hampstead Coach House Extension

Architect: Jonathan Clark
Architects
Location: London, UK
Date of construction: 2004
Photography: Lisa Linder,
Jonathan Clark

The renovation of this former coaching inn is located in the Frognal Conservation area in northwest London. To gain consent from the planners, the side of the existing house was underpinned and the levels dropped by nearly three feet in order to squeeze two storeys below the existing high level eaves.

The owner decided to be bold
with color in her new kitchen.
To prevent an overpowering
feeling, the red was limited to
the top cupboards.

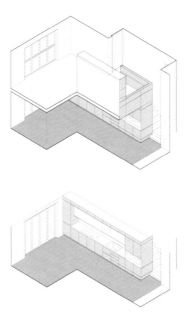

Isometric drawing

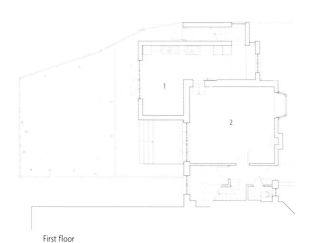

First floor

Upper floor extension

1. Kitchen/dining room
2. Living room
3. Studio

In a small space, space-saving features are the best option, like built-in-cupboards, for example.

Alley Retreat

A former two-storey dilapidated, detached garage set off of a mid-block alley, which housed parking at the upper (street) level and an uninhabitable storage space at the lower (garden) level, was redefined as an alternative entry to the property, integrated with the functions of the main house and turned into a room for guests, children or outdoor entertaining.

Architect: Cary Bernstein
Location: San Francisco,
CA, USA
Date of construction: April 2007
Photography: Sharon Risedorph

A new stair, covered entirely by a large skylight and lined in Ipe siding, connects the garage to the garden below.

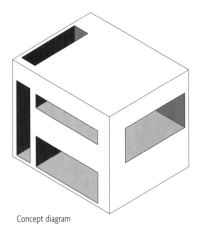

Concept diagram

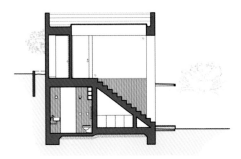

Longitudinal section

Transversal section

Site plan

The structure's simple form is
clad in painted wood siding,
which complements the
neighborhood's construction.

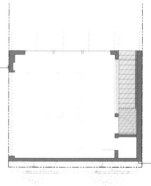

Garage plan

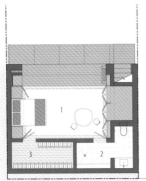

Garden level plan

1. Bedroom
2. Bathroom
3. Wine closet

The former storage area was renovated to create a flexible room; custom cabinetry contains a Murphy bed and ample storage, while a wine closet and washroom are tucked behind the chalkboard slate wall.

The renovation of a 2,600 square-foot mid-century hillside house involved the removal of four interior walls, the raising of a portion of the roof, the changing of the ceiling plane and the demolition of existing interior finishes. The result is a house with more natural light and extended views of the canyon and its indigenous wildlife.

Benedict Canyon Residence

Architect: John Enright, Margaret Griffin, Norma Chung/Griffin Enright Architects
Location: Beverly Hills, CA, USA
Date of construction: 2005
Photography: Benny Chan/ Fotoworks

A portion of the original gabled roof was replaced with an extended plane of the same angle, resulting in a continuous, upwardly-sweeping surface.

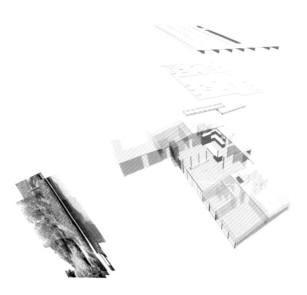

Axonometry

Glass elevation

This new loft-like space
defines discrete functional
areas through changes in the
ceiling plane, placement of
furniture, custom built-ins and
lighting.

An open shelf with slender steel columns replaces the wall that originally separated the family room from the dining room.

1. Family room
2. Living room
3. Dining room
4. Kitchen
5. Master bedroom
6. Bedroom
7. Bathroom
8. Closet
9. Laundry
10. Garage
11. Pool

Floor plan

Orr House

The brief called for an addition and remodel to a 5,080 square-foot 1970s stucco-clad two-storey home built on a steep down-slope lot in semi-rural Saratoga. The architects sensitively combined new and old to create a new, fresh and unique design which is both beautiful to look at and beautiful to live in.

Architect: Swatt Architects
Location: Saratoga, CA, USA
Date of construction: 2005
Photography: César Rubio

The first strategy in this
remodel was to provide
natural light by inserting an
atrium into the center of the
building.

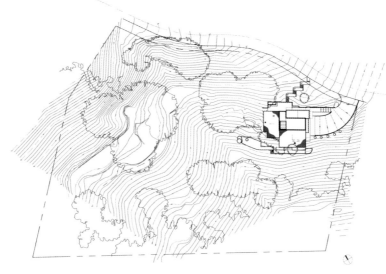

Site plan

1. Garage
2. Laundry
3. Closet
4. Bathroom
5. Bedroom
6. Study
7. Entrance
8. Kitchen/dining
9. Living
10. Master bedroom
11. Master bathroom
12. WC

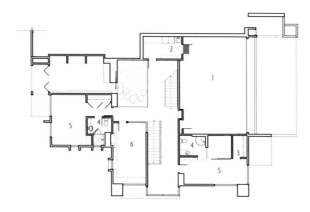

Lower level

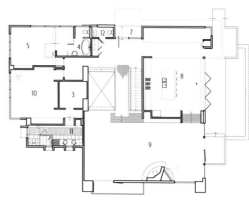

Upper level

The new design introduces a series of overlapping horizontal cedar-clad planes, which serve to extend interior space to the exterior.

The kitchen/dining and living areas share
a new stone terrace edged with a
cantilevered reflecting pool.

The kitchen is planned as a
large multi-purpose space
which includes dining and
flows into the living area.

The owners of this house situated on the shore of a lake asked the architects to superimpose the existing structure with a new architecture. They requested a residential component for two with a private and autonomous space for guests, high ceilings that reference the *bel étage* typical of northern countries and they also expressed a desire to work with color.

Architect: Atelier Oï
Location: Saint-Aubin-Sauges, Switzerland
Date of construction: 2007
Photography: Yves André

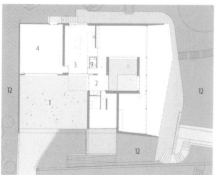

Ground level

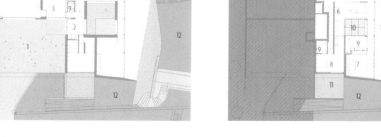

Lower level

1. Entrance yard
2. Hall
3. Studio
4. Garage
5. Living room/dining room
6. Kitchen
7. Bedroom
8. Guest room
9. Bathroom
10. Patio
11. Terrace
12. Garden

Two patios provide a transition between the interior and exterior spaces and act like lanterns, injecting natural light inside.

Color is an important
architectural feature in this
project and interacts with the
earth-based elements.

85

The domestic program is organized in a succession of planes facing the lake.

Mafra House

Architect: A-LDK
Location: Sobreiro Mafra,
Portugal
Date of construction: 2005
Photography: FG+SG
Fotografia de Arquitectura,
Fernando Guerra

This small typical house some 12 miles from Lisbon was in an advanced state of degradation and did not present the minimum conditions of habitability. The interiors and roof were demolished. A structural solution of beams and iron pillars mounted 4 inches from the existing walls made it possible to support the new roof as well as to create a new upper floor. Internally the space was divided into two small independent habitation units, suitable for two small families.

North elevation

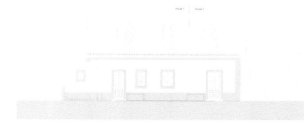

South elevation

East elevation

West elevation

86

The interior volume has a loft-like organization, which allows total freedom of movement within.

HOUSE 1

House 1:
1. Living room
2. Kitchen
3. Void to living room
4. Office
5. Bedroom

House 2:
a. Living room
b. Kitchen
c. Bedroom
d. Terrace

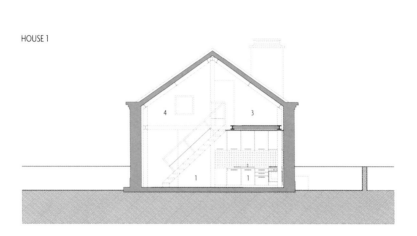

Transversal section

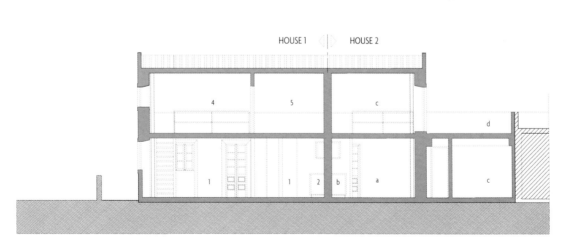

HOUSE 1   HOUSE 2

Longitudinal section

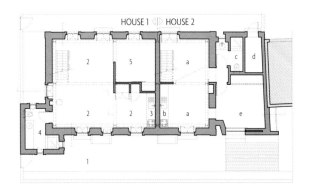

First floor

House 1:
1. South garden
2. Living room
3. Kitchen
4. Bathroom
5. Bedroom
6. Void to living room
7. Office

House 2:
a. Living room
b. Kitchen
c. Bathroom
d. Storage
e. Bedroom
f. Void to living room
g. Terrace

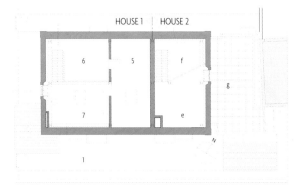

Second floor

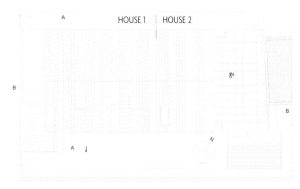

Roof

Private areas are reduced in
size to make common areas
more spacious. Natural light
comes in through several
openings.

88

Color is an important feature of the interior design as it helps define different domestic functions in an open layout.

The traditional exterior image
of this typical Portuguese
house contrasts greatly with
the contemporary open-plan
layout and modern furnishings
within.

Original

This house is unique in the sense that it was built with the owner's family structure in mind: a single-father household. Winner of a 2005 American Institute of Architects Minnesota Honor Award, the linear open plan sits between an easterly pond and a westerly wetland. The house represents its occupants (father and son) in several ways. A black, monolithic masonry base contains the main domestic areas, while two white bedroom boxes sit on top of this base.

Streeter House

Architect: David Salmela/
Salmela Architect
Location: Deephaven, MN, USA
Date of construction: 2005
Photography: Peter Bastianelli
Kerze

Site plan

This house is a unified
statement of sustainable
artistry: custom-made black
12-by-12-by-24-foot concrete
block, structural insulating
panels, wood laminated
structure, slatted wood
screens and recycled cypress.

1. Main entrance
2. Courtyard
3. Garage
4. Equipment
5. Screen porch
6. Living room
7. Dining room
8. Kitchen
9. Laundry room/bar
10. Office
11. Powder room
12. Guest bedroom/
    sauna lounge
13. Bathroom
14. Sauna
15. Bedroom
16. Deck

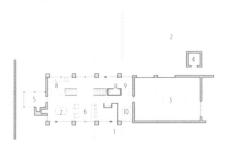

First floor

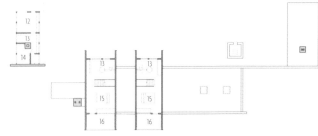

Second floor

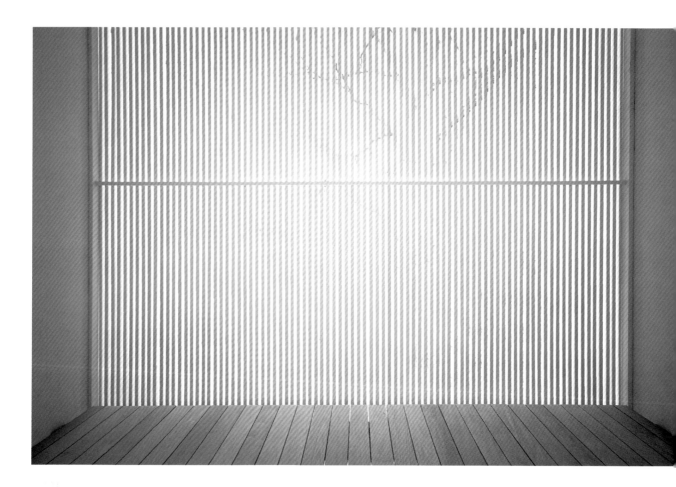

With only two interior doors,
this is a far cry from any
conventional house, rather
every aspect of the interior
has been designed to suit the
way a father and son would
want to live.

92

The garage takes up almost half of the house's main level, with plenty of room for motorcycles, equipment and tool storage.

T-Stomach

Architect: Naoki
Terada/Teradadesign Architects
Location: Saitama, Japan
Date of construction: 2006
Photography: Yuki Omori

The brief for this new house read: "We would like to live in a house where the whole family can always be together in one big room." The architects based their design on the human digestive organs, where the functions are separated by the narrowing and winding of the organs, neatly folded inside the torso. This house is one big winding tube, which widens and narrows as it leads from the corridor and living room to the bedrooms.

North elevation

South elevation

Floor plan

1. Entrance
2. Public space
3. Private space 1
4. Private space 2
5. Guest room
6. Courtyard/pond

To emphasize the shape and expansion of the space, a variation in darkness of color is created by a saturation of light.

Full gloss epoxy resin paint
was used for the floors and
urethane resin paint for the
walls and ceilings. The color
gradually fades as you stare
at it.

The winding walls moderately block the view to maintain distance between family members, or by following the curves, one discovers different sceneries.

The client of this dwelling requested "to be able to see green from every room." The architects' previously applied concept "interface" was perfect for this 3875 square-foot lot surrounded by houses on both sides. Made of a peripheral screen with a green zone behind it, this concept is a combination which acts as a filter between the inside and outside of a house.

House in Shimogamo

Architect: Edward Suzuki
Associates
Location: Yakoucho,
Shimogamo, Kyoto City, Japan
Date of construction: 2006
Photography: Yasuhiro
Nukamura

Warm, natural, smoked bamboo louvers contrast with the cold, high-tech expression of the circular screen of frosted glass.

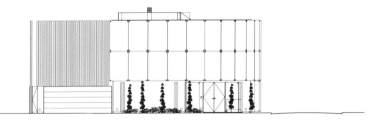

East elevation

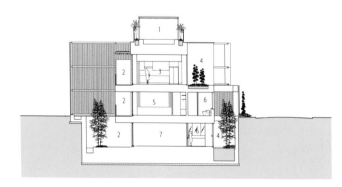

1. Roof terrace
2. Terrace
3. Kitchen
4. Patio
5. Master bedroom
6. WC
7. Play room

One of the screens applied to this house is a circular, frosted glass screen, which envelops the second storey on the northeast.

Transversal section

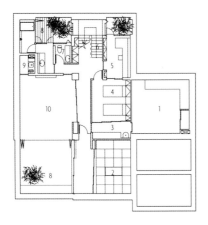

Basement

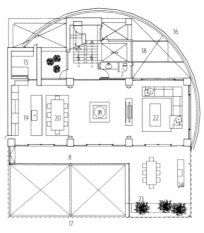

Second floor

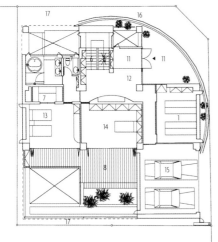

First floor

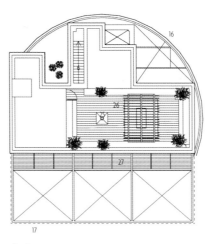

Roof

1. Storage
2. Japanese room
3. Utility room
4. Guest room
5. Study
6. Stairs
7. WC
8. Terrace
9. Laundry room
10. Play room
11. Entrance
12. Foyer
13. Master bedroom
14. Bedroom
15. Garage
16. Glass screen
17. Bamboo screen
18. Glass canopy
19. Kitchen
20. Dining room
21. Fireplace
22. Living room
23. Bench
24. BBQ
25. Chimney
26. Roof terrace
27. Louvered glass canopy

A vertically-louvered, smoked
bamboo screen offers some
level of privacy, while
retaining the silhouette of the
houses next door.

Bamboo louvers offer a
degree of privacy respecting
the surrounding landscape by
filtering it but not totally
closing it off.

The interiors are kept simple
with neutral colors, thus
creating a relaxed and
peaceful environment.

A Japanese room in the basement flanks a sunken patio along with a study and family room.

Ecker Abu Zahra
Honeyhouse

Built for part-time apiculturists, this simple cuboid structure with a quadrate 49-by-49-foot floor plan is a tranquil spot in the woods. The house is covered in copper-plate, a material which changes color from gloomy gold to a green-blue dull surface which blends in with the color of the woods. This material also shields radiation and thus creates a peaceful inner space from which to observe nature.

Architect: Hertl Architekten
Location: Luftenberg, Austria
Date of construction: July 2006
Photography: Paull Ott-Graz

South elevation

West elevation

North elevation

East elevation

As the sun sets, the materiality of the copper changes from a gloomy gold to a green-blue dull surface which fuses with the color of the woods.

To underline the simple inner structure of this house, beige was chosen for the surface of the walls, ceilings and floors, thus creating a silent contrast to the openings.

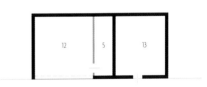

Garage plan

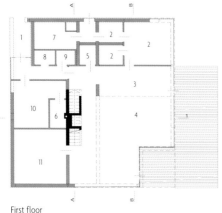

First floor

Second floor

1. Entrance
2. Apiary
3. Kitchen
4. Hall (representation & living)
5. Storage
6. Bathroom
7. Technical equipment
8. Wardrobe
9. Toilet
10. Guest room
11. Library
12. Garage
13. Cellar
14. Sauna
15. Bedroom
16. Dressing room
17. Roof terrace
18. Office
19. Void

Villa Room

Architect: Architectenbureau
Paul de Ruiter, Rob Hootsmans
Location: Rehnen,
The Netherlands
Date of construction: 2004
Photography: Pieter Kers,
Rob't Hart

The main concern in the design of this home was to allow the daily activities of the residents to form the main structure of the building. One of the owners needed a studio, a view of the garden and an exhibition area, while the other wanted a house that could be controlled in relation to both the indoor and the outdoor climate and to the degree of privacy. The result is a villa which does not have a traditional living room.

The villa is arranged around an open space with a roof through which light can enter. The northern light falls on the center of the house and is ideal for the studio on the ground floor.

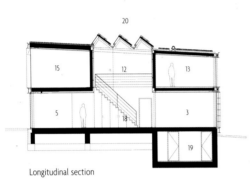

Longitudinal section

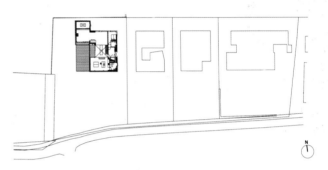

Site plan

First floor

Second floor

1. Entrance hall
2. High space
3. Kitchen-dining room
4. Back kitchen
5. Studio
6. Living room
7. Storage room
8. Spare room
9. Bathroom
10. Loggia
11. Terrace
12. Landing/working space
13. Study
14. Void
15. Bedroom
16. Bathroom
17. Walk-in closet
18. Atelier
19. Cellar
20. Sun panels

In contrast to the light
interiors, the façade has a
heavy and robust character.
The open and closed sections
in the façade provide a wide
variety of views of the
surrounding landscape.

The villa is fitted with domotic systems such as electrically-operated aluminum sun blinds which keep out sunlight and ensure privacy.

Floating Cube

Architect: Hitoshi Saruta/
CUBO Design Architect
Location: Fujisawa City,
Kanagawa Prefecture, Japan
Date of construction: April 2005
Photography: Yasuno Sakata

The client of this house on a small hill facing the Pacific Ocean had three main requests: an outstanding structure which passers-by would admire; a space with a sense of transparency; and a variation in the concept embodied by each floor. The architect proposed a structure resembling a floating box with a protruding first floor.

The first floor of this "floating box" has been expanded and extensively protrudes forward, thus the second floor is smaller in size and causes less loss of sunlight to the building's northern neighbor.

As there is little sunlight on the ground floor, this has been reserved for the bedroom area. The interior design of this floor is based on a Balinese resort.

1. Entrance
2. Closet
3. Master bedroom
4. Bedroom
5. Kitchen
6. Dining room
7. Living room
8. Deck
9. Bathroom

First floor

Second floor

Third floor

The design of the first floor is "New York Soho;" a modern space made up of a unified monochrome kitchen and living area.

One of eleven houses of the Commune by the Great Wall, each one designed by an Asian architect, Suitcase House sits at the foot of the Great Wall, at the head of the Nangou Valley. This dwelling is made up of a stacking of strata. The middle stratum is a place for habitation, activity and flow, while the bottom stratum acts as a container for dedicated spaces with compartments concealed by floor panels.

Architect: Gary Chang, Andrew Holt, Howard Chang, Popeye Tsang, Yee Lee
Location: Beijing, China
Date of construction: 2004
Photography: EDGE Design Institute

Middle level

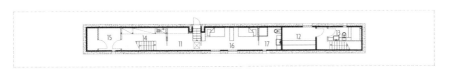

Lower level

1. Living room
2. Dining room
3. Bedroom
4. Storage
5. Bathing
6. Study
7. Kitchen
8. Cloak room
9. Meditation
10. Audio-visual
11. Library
12. Sauna
13. Laundry room
14. Pantry
15. Boiler room
16. Butler's bedroom
17. Butler's bathroom

109

An exercise in rethinking the nature of intimacy, privacy, spontaneity and flexibility, the inner layer of this house consists of a series of screens, which form a matrix of openings.

110

The middle level is a metamorphic volume which slides effortlessly into an open space to a sequence of rooms.

Glass Box

Architect: Gwenael Nicolas/
Curiosity
Location: Shibuya, Tokyo, Japan
Date of construction: 2005
Photography: Daici Ano

This project was designed as a product before the land was even found. This glass house surrounded by a walkway/gallery which connects the floors is not defined by the walls and floor, but by the movement of the user within the space. The interior is defined by a series of scenes reflecting how the user will appear and disappear from floor to floor.

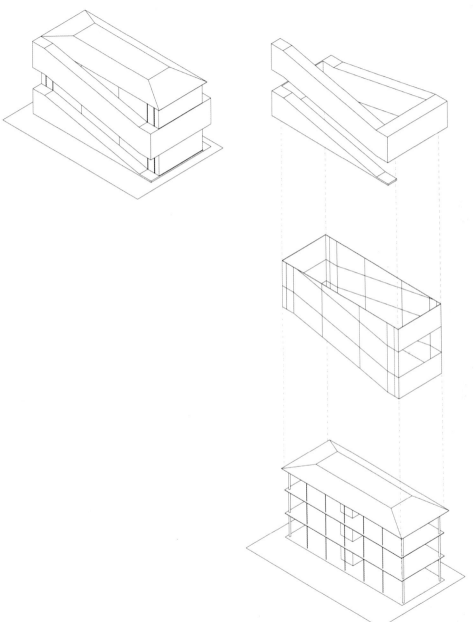

Axonometry

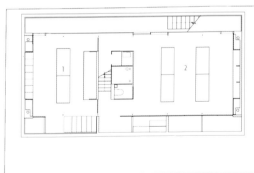

Basement

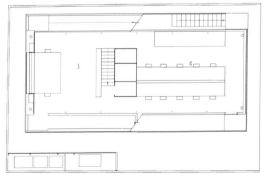

First floor

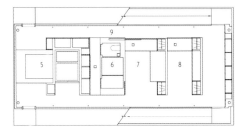

Second floor

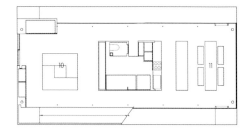

Third floor

1. Workshop
2. Meeting space & library
3. Personal working space
4. Staff working space
5. Master bedroom
6. Bathroom
7. Children's room
8. Guest room
9. Corridor
10. Living space
11. Kitchen/dining space

111

As the house is surrounded
by a walkway, the interior is
designed in three dimensions,
so that everything is visible
from floor to ceiling level.

The size, proportion, height
and materials of the furniture
in this house makes one
re-evaluate the perception of
scale.

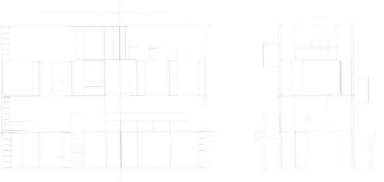

Longitudinal section

Transversal section

For the architects, this house
defines life; each action and
movement is defined and
controlled by the design of
the space and creates the
balance of life.

Ö House

This single-family house is designed on a 46-by-203-foot site and offers 26-by-98-foot of living pleasure. The folded roofscape is a response to the height variations of the covered program, including a garage and storage room, master bedroom, entrance area, a small guest bathroom and a living room and kitchen with a 14.7 foot height and panoramic views over the garden.

Architect: Share Architects
Location: St. Pölten, Austria
Date of construction: 2004-2005
Photography: Franz Ebner

Site plan

South elevation

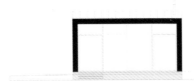

West elevation

The shape of the house is
determined by the different
room heights, which also
reflects in the cost of running
this house.

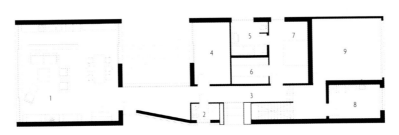

1. Living room/kitchen
2. Toilet/shower
3. Corridor
4. Studio
5. Bathroom
6. Wardrobe
7. Master bedroom
8. Laundry room
9. Garage

Floor plan

The abundant use of glass
ensures fluidity, transparency
and connectivity throughout
the entire house.

At 13 feet in height, the living room enjoys the highest part of this house, which is defined by the domestic programs.

117

Thanks to the rectangular
floor plan, each room of this
house enjoys panoramic
views over the garden.

This project required varying and multiple spaces within a limited allowable building area. The building makes a reference to the "cranks" found in the surrounding Edwardian roofs, terracotta tiles and brick chimneys. The building rotates off the orthogonal axis of the side boundaries to address the rear (angled) boundary, which is marked by continuous vertical "tears."

St. Kilda House

Architect: Saaj Design
Location: St. Kilda, Victoria, Australia
Date of construction: 2006
Photography: Patrick Redmond

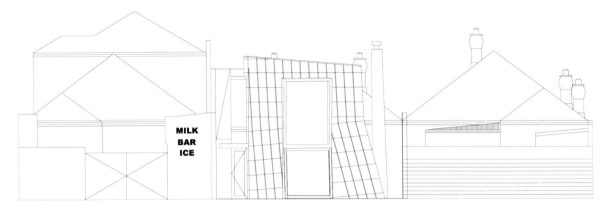

Southwest elevation

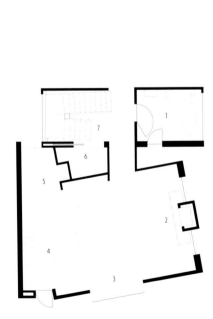

First floor

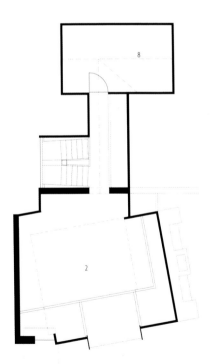

Second floor

1. Bathroom
2. Living room
3. Dining room
4. Kitchen
5. Pantry
6. Laundry room
7. Staircase
8. Bedroom/study
9. Second living room

The clean straight lines and
clear colors used throughout
create relaxing and
sophisticated interiors.

The scale of the mezzanine
level with the second living
room is more intimate than
the open-plan lower level.

A mirror placed along one side of the
room makes this bathroom look twice
its size.

Tattoo House

Architect: Andrew Maynard
Architects
Location: Fitzroy North, Victoria,
Australia
Date of construction: 2007
Photography: Peter Bennetts

The brief for this project was to create a new living and kitchen space for a growing young family and to create an open plan with plenty of natural light and high ceilings. Keeping in mind the dual requirements of a council overlooking regulations and glare reduction, the architects introduced super-graphic UV-stable tree stickers. The pattern creates a playful series of shadows across the interior spaces.

1. Kitchen
2. Dining room
3. Living room
4. Bedroom
5. Bathroom
6. Laundry room
7. Deck
8. Tank

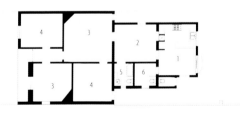

Original floor plan

New floor plan with addition

To comply with the legislation which dictates a 75 % opacity to second-storey windows, UV-stable stickers were applied to the exterior of this extension.

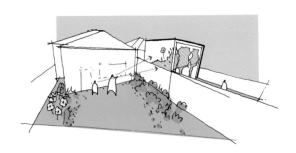

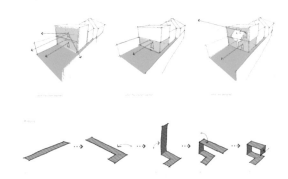

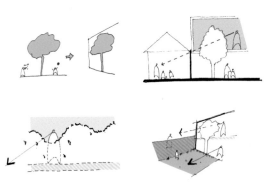

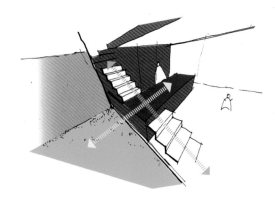

Diagrams

Maison A

Faced with a very tight budget and urban building constrictions, the architects decided to renovate the original structure and add a light structure on top of the existing garage to extend the house. Different materials are combined to create a patchwork of existing (brick and stonework painted white) and added materials (white steel roof tiling and translucid panels).

Architect: Chartier-Corbasson
Architectes
Location: Paris, France
Date of construction: 2004
Photography: Emmanuel Barbe

Working with a very tight
budget and building
restrictions, the architects
added a light structure to
the existing garage as an
extension to the existing
house.

Transversal section

Longitudinal section

1. Garage
2. Kitchen/living/dining

122

A series of windows placed laterally and near the zenith allows light to flood the interior of the volume housing the kitchen and living areas.

Architect: Hitoshi Sarut/CUBO
Design Architect
Location: Matsumoto City,
Nagano Pref, Japan
Date of construction: March
2007
Photography: Yasuno Sakata

Considering cost limitations and unique site challenges, this small house built on
land which was originally a parking area was designed in cooperation with the client.
Both the interior and exterior emit a feeling of calm. The exterior of the house is
formed minimally with angled walls. Inside, walnut floors and white walls provide a
wonderful contrast.

1. Bedroom
2. Bathroom
3. WC
4. Laundry room
5. Living room
6. Dining room
7. Kitchen

First floor

Second floor

123

First priority is given to
the creation of convenient
access to all facilities for all
the inhabitants. The two
bedrooms are placed on the
ground floor to ensure
privacy.

The continuity of space is
ensured by making the
division of spaces indistinct
through the use of Japanese-
style lattice so the presence
of other family can be sensed.

125

Designed to eventually become a guest house, this dwelling includes a parent's mezzanine room, minimal bathroom, bench kitchen and a long fixed couch, which doubles as the children's toe-to-toe bed.

The shape of this house
follows the minimum head
clearance to allow a
mezzanine level, thus
providing maximum space at
minimum cost.

127

This is a very economical
house to run, with heating
costs kept to an absolute
minimum thanks to the
weekend home's small
volume.

Spectacular

Villa NM

This is a house for summers, for weekends and for stolen time. Situated on top of a hill, the house affords panoramic views over the surrounding landscape, which can be enjoyed from within thanks to walls which consist of floor-to-ceiling windows, tinted like the sky at dusk. The exterior of the house is colored like the earth while the cool white flowing spaces inside make you part of the constantly changing landscape around you.

Architect: UN Studio
Location: Upstate New York, USA
Date of construction: 2006
Photography: Christian Richters

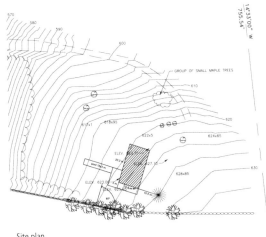

Site plan

View from north

View from south-east

View from west

GEOMETRY OF THE FIVE TWISTED ELEMENTS

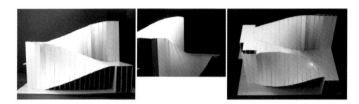

Combination of twists

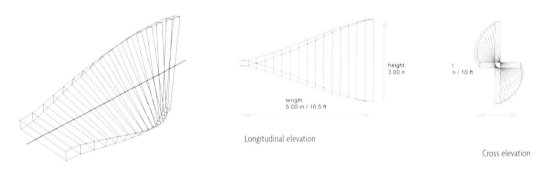

Axonometric view

Longitudinal elevation

height
3.00 n

length
5.00 m / 16.5 ft

t
n / 10 ft

Cross elevation

width
1.50 m /

Plan

128

The kitchen and dining area on the ground floor are connected by a ramp to the living space above. This height change allows for a sweeping outlook over the surrounding scenery.

Casa Nuria Amat

The main challenge for Jordi Garcès was how to build a house on a cliff without damaging the cliff but making the most of it at the same time. The rock is the place's basic material and is present everywhere in the house. The house is made up of two prisms which cross at a precise point, protecting the emptiness of the patio and creating a double-height interior space.

Architect: Jordi Garcès
Location: Tamariu, Spain
Date of construction: 2006
Photography: Jordi Miralles

Situated on a cliff overlooking the cape of
Begur on Spain's Costa Brava, this stunning
house has been built into the cliff.

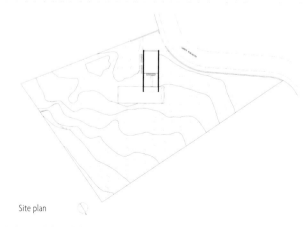

Site plan

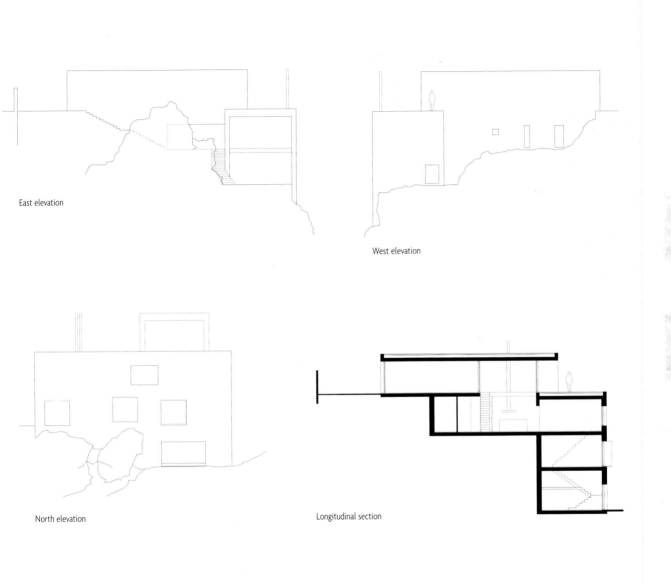

East elevation

West elevation

North elevation

Longitudinal section

Hidden in the rocks, the
exterior walls of the volumes
which make up this luxurious
seaside home have been
plastered and painted silver.

A pool has been carefully carved into the rock, without damaging the cliff it sits on, blending in with its surroundings.

The house is made up of two bodies. The point of intersection of these two perpendicular blocks creates a double-height interior space.

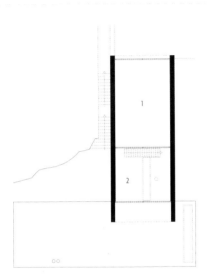

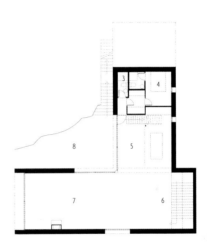

1. Multi-use space
2. Void
3. Courtesy bathroom
4. Service quarters
5. Kitchen
6. Living room
7. Dining room
8. Patio
9. Bedroom
10. Dressing room
11. Bathroom
12. Guest suite
13. Guest terrace

First floor

Basement level 1

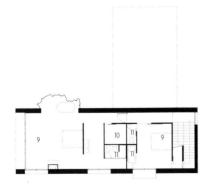

Basement level 2

Basement level 3

The double-height kitchen and dining area is both a resting area and a passing-through area with views towards the horizon.

The living, dining and kitchen areas are clearly defined spaces, which flow both horizontally and vertically.

The geometry of black and
white creates a calm
and luxurious atmosphere in
one of the house's bedrooms.

133

Modern furnishings and minimalist colors both contrast and complement the warm color of the rock.

Holmes House

This house sits on a rocky formation and has privileged views of the bay, the open sea and the mountains. The basic concepts of the design were to make the most of the views and the rocks and to create transparency, fusing the interior spaces with the exterior spaces. Thanks to a base structure of concrete columns and lots of glass, the architect has managed to blend the house into its surroundings.

Architect: Victor Cañas
Location: Playas del Coco,
Costa Rica
Date of construction: 2004
Photography: Jordi Miralles

To eliminate the need for supporting walls and to create transparency and integration between the interior and exterior, the structure is based on concrete columns supporting light roofs.

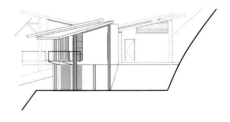

Entrance façade

The narrow and elongated shape of the plot has led to a lineal configuration of spaces, organized alongside a longitudinal corridor, which runs the length of the house.

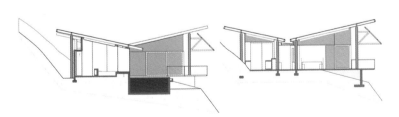

Transversal section master bedroom   Transversal section kitchen and dining

One of the main requirements of this
project was to make the most
of the views.

Basement (apartment)

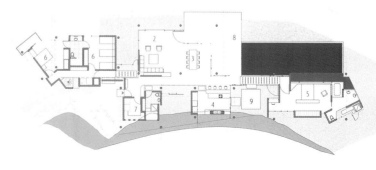

First floor

1. Apartment
2. Living room
3. Dining room
4. Kitchen
5. Master bedroom
6. Guest bedroom
7. Laundry room
8. Terrace
9. Breakfast terrac

To offset the excessive
sunlight that the main living
area receives, aluminum sun
breakers protect the space
and visually enlarge it as the
area is reflected back to
the interior.

The social areas are placed at the center
of the layout with the bedrooms and
guest apartment on one side and the
master bedroom and pool on the other.

The bed is placed opposite a
large window facing the open
sea ahead. This area enjoys a
calming atmosphere as well
as refreshing breezes from
the sea.

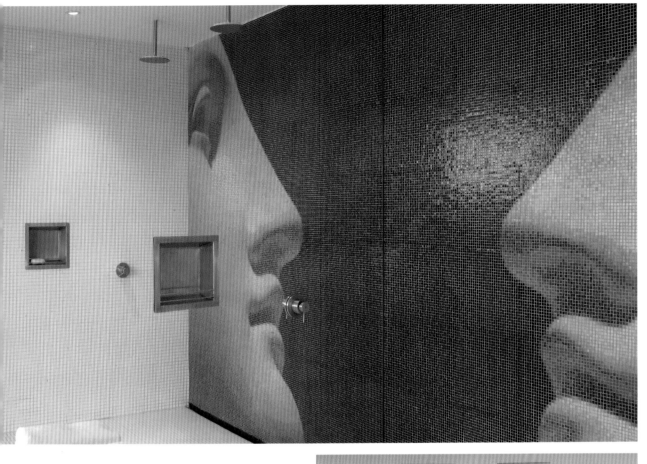

Villa Soravia

Because of building regulations, the form of this vacation home on the shore of Lake Millstatt is defined by the original gable roof, a slanted tower, a generously defined exterior and an inimitable spatial structure. To elevate the upper area of the former house to create a free floor plan on the ground level, a "table" platform of reinforced concrete is introduced.

Architect: Coop Himmelblau
Location: Millstatt, Austria
Date of construction: 2006
Photography: Gerald Zugmann

Site plan

A terrace extends the living room out toward the lake. This room's ceiling-high glass panels can be opened, thus allowing inside and outside spaces to flow together.

The living room features a
Sicilian olive tree to reinforce
the Mediterranean flair of a
summer villa and a bespoke
wine rack which can ascend
from the wine cellar into the
living room when needed.

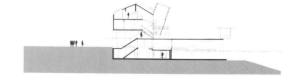

Longitudinal section through the stair tower and pier

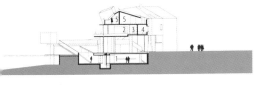

Transversal section through the living room, master bedroom and wine cellar

Transversal section through the master bedroom and wine cellar

1. Wine cellar
2. Bedroom
3. Guest bathroom
4. Guest bedroom
5. Room
6. Living room
7. Hammam
8. Bathroom

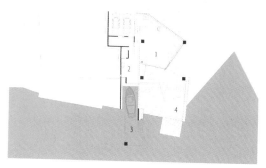

First floor

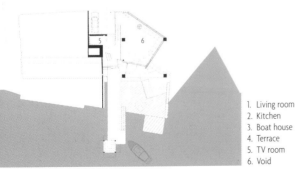

1. Living room
2. Kitchen
3. Boat house
4. Terrace
5. TV room
6. Void

Second floor

The lower level contains the business, storage and technical spaces as well as a mahogany-paneled wine cellar with a bar and ample seating for tastings.

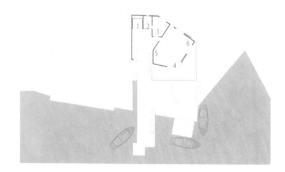

Third floor

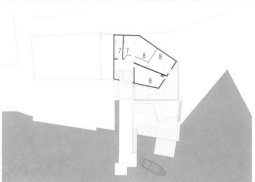

Fourth floor

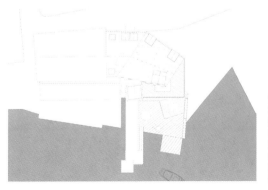

Roof

1. Sauna
2. WC
3. Guest bedroom
4. Master bedroom
5. Dressing room
6. Master bathroom
7. Bathroom
8. Bedroom

Contemporary Villa

Architect: Steven Haas
(architect), Robert Nassar, Lisa
Tillinghast (interior designers)
Location: Litchfield County,
CT, USA
Date of construction: 2007
Photography: Peter Aaron/ESTO

The design of this country home in Connecticut reflects the owners' passion for Tuscany, defined by the ochre-colored stucco wall typical of Tuscan villas. A wall divides the landscape into two dramatically different settings: the public side surrounded by a typical New England wooded environment and the private Tuscan side.

The living room has a very large and imposing volume, which has been "humanized" by introducing oversized sofas and tables.

The media room is an informal space which seems to flow into the room beyond. Custom millwork houses the television and audio visual equipment.

1. Entry terrace
2. Gallery
3. Kitchen/family room
4. Screened porch/terrace
5. Dining room
6. Living room
7. Master suite
8. Gym
9. Study
10. Media room
11. Conservatory
12. Garage
13. Guest house
14. Pool/terrace

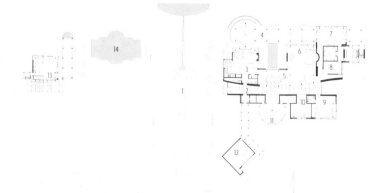

Floor plan

The family room is centered around an
open hearth and is adjacent to the
country-style kitchen topped with inlaid
marble.

The guest house, with the living area on the upper level and the bedroom suite on the lower level, has an open loft-like plan.

Soft colors, silky textures,
gently curved furniture and
luminous materials give the
master bedroom a romantic
feeling.

The master bathroom has an Asian atmosphere. A bamboo canopy construction screens the luminous ceiling above.

Architect: Akira Yoneda/
Architecton, Masahiro Ikeda
Location: Kodaira, Tokyo, Japan
Date of construction: 2006
Photography: Koji Okumura

Designed for a young manga artist, this 6,000 square-foot house located in Tokyo's dense urban district was inspired by the client's futuristic cartoons. A stack of boxy volumes creates dramatic cantilevers. Each volume contains a different domestic area with the living spaces on top, a studio below grade and a white garage for the artist's collection of cars.

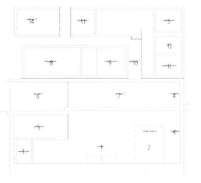

Longitudinal section

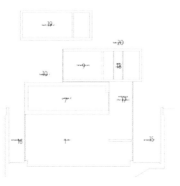

Transversal section

1. Studio
2. Work room 2
3. Washroom
4. Machine
5. Playroom
6. Garage 1
7. Garage 2
8. Living room
9. Tatami room
10. Balcony 1

11. Reading room
12. Loft
13. Bedroom 1
14. Balcony 3
15. Areaway 1
16. Areaway 2
17. Entrance (for studio)
18. Pet room
19. Bedroom 2
20. Balcony 3

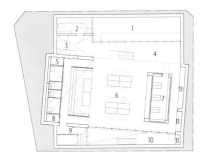

Basement

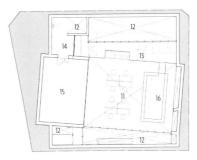

Middle basement

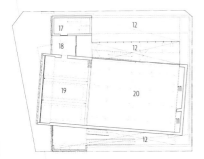

First floor

Second floor

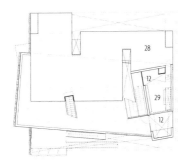

Middle second floor

Third floor

1. Areaway 1
2. Work room 1
3. Water supply
4. Lounge 1
5. Stock room
6. Studio
7. Book storage
8. Shower room
9. Sleeping beds
10. Areaway 2
11. Machine
12. Void
13. Lounge 2
14. Storage
15. Playroom
16. Work room 1
17. Entry to residence
18. Entry to studio
19. Garage 1
20. Garage 2
21. Dining room/kitchen
22. Pet room
23. Living room
24. Tatami room
25. Reading room
26. Sunroom
27. Balcony 1
28. Balcony 2
29. Loft
30. Balcony 3
31. Master bedroom
32. Bedroom 1
33. Corridor
34. Bedroom 2
35. Washroom
36. Sauna
37. Bathroom

This house is situated on a 13 foot-wide street, which could be expanded to 82 feet. The structure is conceived as an index for an intermediate view linking distant and close views.

The overall structure is linear on the outside. The plan and section are given circulatory properties to achieve a balance between the object's independence and the latitude of human activity.

Residence Klosterneuburg

This spacious mansion for a young family living north of Vienna is situated on a slope overlooking the valley below. The base of the building roughly follows the incline of the slope—integrated terrace levels on the exterior meet the character of the slope and the surrounding landscape—while an upper, more compact part of the building seems to rise from the site.

Architect: Andreas Schmitzer
Location: Klosterneuburg,
Vienna, Austria
Date of construction: 2006
Photography: Nadine Blanchard

The goal of the design on this spacious mansion overlooking the valley below was to give the impression of a completely white box, including the top, visible from above.

Site plan

The ground floor consists of several levels which follow the downward slope. These levels divide the lavish loft-like living space into various functional areas.

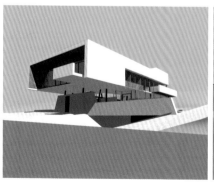

West elevation

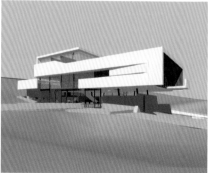

East elevation

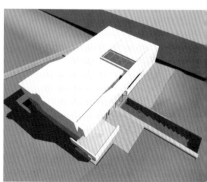

Top view, north

op view, south

146

A cantilevered structure situated above the ground floor houses the private rooms and a rooftop terrace from which to enjoy spectacular views over the valley.

Basement

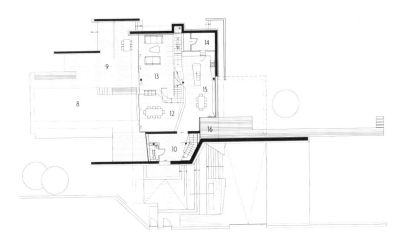

First floor

Second floor

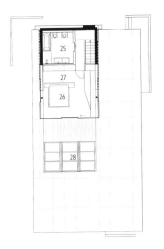

Third floor

1. Garage
2. Separator
3. Storage room
4. Fitness room
5. Rest area with steam bath
6. Technical room
7. Laundry room
8. Pool
9. Terrace
10. Entrance/hallway
11. WC
12. Dining room
13. Living room
14. Larder
15. Kitchen
16. Breakfast terrace
17. Void
18. Studio
19. Balcony
20. Library/gallery
21. Playing area
22. Children's room
23. Guest room
24. Bathroom
25. Master bathroom
26. Master bedroom
27. Wardrobe
28. Rooftop terrace

Almost attached to the building, the pool
becomes a part of the house thanks to
large sliding doors, which allow
a seamless connection between the
interior and exterior.

Necklace House

The architects created a C-shaped wall surrounding the open living area so the client can enjoy the integration with the garden without any need for. Because the area is prone to heavy snowfall—over 6.5 feet high at times—rooms were kept as far away from the ground as possible. Detached rooms are spread along the wall in the shape of an annex so that the beautiful surrounding mountain chain can be enjoyed from each room.

Architect: Hiroshi Nakamura/
NAP Architects
Location: Yamagata, Japan
Date of construction: September
2006
Photography: Daici Ano

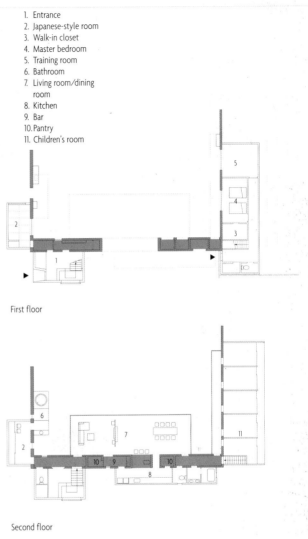

1. Entrance
2. Japanese-style room
3. Walk-in closet
4. Master bedroom
5. Training room
6. Bathroom
7. Living room/dining room
8. Kitchen
9. Bar
10. Pantry
11. Children's room

First floor

Second floor

147

This house is built as a series of rooms strung together like a necklace, defying the conventional method of layouts.

Each room feels like a detached box in itself, thus offering a maximum amount of privacy for the dwellers.

8

The garden and surrounding landscape are integrated into the living room. Each detached room is spread along a structural wall so beautiful views can be enjoyed all around.

Though each room is detached, the inhabitants can move around the house easily without having to go outside.

The reflection of light created
in the bathroom by the
sixteen-hundred windows
provides a unique bathing
experience.

Directory

Aidan Halloran/ITN Architects
2/184 Brunswick Street
Fitzroy, Victoria 3065
Australia
T. +61 3 9416 3883
F. +61 3 9416 3847
itn@itnarchitects.com
www.itnarchitects.com

Akira Yoneda/Architecton
1-7-16-612 Hon-cho Shibuya
Tokyo 151-0071
Japan
T. +81 3 3374 0846
F. +81 3 5365 2216
a-tecton@pj8.so-net.ne.jp
www.architecton.co.jp

A-LDK
Cristovao Rodrigues de Acenheiro 6-3 esq.
1300-151 Lisbon
Portugal
T. +351 21 3644369
F. +351 91 4080177
daiji.kondo@mail.telepac.pt

Alonso Balaguer & Arquitectos Asociados
Bac de Roda 40
08019 Barcelona
Spain
T. +34 93 303 41 60
F. +34 93 303 41 61
estudio@alonsobalaguer.com
www.alonsobalaguer.com

Andreas Schmitzer
Mariahilfer Street 101/3/1/48
1060 Vienna
Austria
T. +43 1 526 88 26
F. +43 1 526 99 91
office@projecta01.com
www.projecta01.com

Andrew Maynard Architects
Level 1, 36 Little Bourke Street
Melbourne, Victoria 3003
Australia
T. +61 3 9654 2523
F. +61 3 8640 0439
info@andrewmaynard.com.au
www.andrewmaynard.com.au

Architectenbureau Paul de Ruiter
Leidsestraat 8-10
1017 PA Amsterdam
The Netherlands
T. +31 20 626 32 44
F. +31 20 623 70 02
info@paulderuiter.nl
www.paulderuiter.nl

Atelier Oï
Signolet 3
2520 La Neuveville
Switzerland
T. +41 32 751 56 66
contact@atelier-oi.ch
www.atelier-oi.ch

Atelier Tekuto
301-6-15-16 Honkomagome
Bunkyo-ku
Tokyo 113-0021
Japan
T. +81 3 5940 2770
F. +81 3 5940 2780
info@tekuto.com
www.tekuto.com

Att Architekten
Bauerngasse 12
90443 Nuremberg
Germany
T. +49 9 1127 44 79-0
F. +49 9 1127 44 79-44
mg@att-architekten.de
www.markus-gentner-architekt.de

Bligh Voller Nield
365 St. Paul's Terrace
Fortitutde Valley, QLD 4006
Australia
T. +61 7 3852 2525
F. +61 7 3852 2544
brisbane@bvn.com.au
www.bvn.com.au

Boyarsky Murphy Architects
64 Oakley Square
London NW1 1NJ
UK
T. +44 207 388 3572
F. +44 207 691 0847
nsb@boyarskymurphy.com
www.boyarskymurphy.com

Cary Bernstein
2325 Third Street, Studio 341
San Francisco, CA 94107
USA
T. +1 415 522 1907
F. +1 415 522 1917
cary@cbstudio.com
www.cbstudio.com

Chartier-Corbasson Architectes
chart-corb@freesurf.fr

Clinton Murray, Shelley Penn/Clinton Murray Architects
PO Box 794 Williamstown
Victoria 3016
Australia
T. +61 3 9397 7624
F. +61 3 9397 7694
sjpenn2@bigpond.net.au
www.clintonmurray.com.au

Coop Himmelblau
Spengergasse 37
1050 Vienna
Austria
T. +43 1 546 60 334
F. +43 1 546 60 600
communication@coop-himmelblau.at
www.coop-himmelblau.at

Danny Fox/Daniel R. Fox Architect
Level 4, Room 204
303 Adelaide Street, Brisbane
Queensland 4000
Australia
T. +61 7 3211 7183
F. +61 7 3211 7183
danfox@optusnet.com.au

David Salmela/Salmela Architect
630 West Fourth Street
Duluth, Minnesota 55806
USA
T. +1 218 724 7517
F. +1 218 728 6805
ddsalmela@charter.net
www.salmelaarchitect.com

Edward Suzuki Architects
1-15-23 Seta, Setagaya-ku
Tokyo
Japan
T. +81 3 3707 5272
F. +81 3 3707 5274
esa@edward.net
www.edward.net

Gary Chang, Andrew Holt, Howard Chang, Popeye Tsang, Yee Lee/Edge Design Institute
Suite 1604, Eastern Harbour Centre
28 Hoi Chak Street, Quarry Bay
Hong Kong
China
T. +85 2 2802 6212
F. +85 2 2802 6213
edgeltd@netvigator.com
www.edgedesign.com.hk

Gwenael Nicolas/Curiosity
2-13-16 Tomigaya Shibuya-ku
Tokyo 151-0063
Japan
T. +81 3 5452 0095
F. +81 3 5454 9691
info@curiosity.jp
www.curiosity.jp

Hertl Architekten ZT KEG
Zwischenbrücken 4
4400 Steyr
Austria
T. +43 7252 46944
F. +43 7252 47363
steyr@hertl-architekten.com
www.hertl-architekten.com

Hiroshi Nakamura/NAP Architects
T. +81 3 3709 7936
F. +81 3 3709 7963
nakamura@nakam.info
www.nakam.info

Hitoshi Saruta/CUBO Design Architect
3-17-20 Hishinuma
Chigasaki City, Kanagawa Pref
Japan
T. +81 467 54 6994
F. +81 467 54 7035
cubo@cubod.com
www.cubod.com

Hulting Arkitekter
Karl Johansgatan 29
414 59 Göteborg
Sweden
T. +46 32 14 30 50
F. +46 708 910 310
peter@hulting.se
www.hulting.se

John Enright, Margaret Griffin, Norma Chung/Griffin Enright Architects
12468 Washington Blvd.
Los Angeles, CA 90066
USA
T. +1 310 391 4484
F. +1 310 391 4495
info@griffinenrightarchitects.com
www.griffinenrightarchitects.com

Jonathan Clark Architects
Second Floor, 34-35 Great Sutton Street
London EC1V 0DX
UK
T. +44 207 608 1111
jonathan@jonathanclarkarchitects.co.uk
www.jonathanclararchitects.co.uk

Jordi Garcès
D'en Quintana 4
08002 Barcelona
Spain
T. +34 93 317 31 88
F. +34 93 317 22 65
jordigarces@jordigarces.com
www.jordigarces.com

Karin Windgårdh/Wingårdh Arkitektkontor
Kungsgatan 10 A
411 19 Göteburg
Sweden
T. +46 31 743 70 54
F. +46 31 711 98 38
karin.wingardh@wingardhs.se
www.wingardhs.se

Lisa Tillinghast/LST Design
80 Central Park West, Suite 19 B
New York, NY 10023
USA
T. +1 212 580 0122
F. +1.212.714 7901
lstdesign@rcn.com
www.haas-tillinghast.com

Lorcan O'Herlihy Architects
5709 Mesmer Avenue
Culver City, CA 90230
USA
T. +1 310 398 0394
F. +1.310.398 2675
loh@loharchitects.com
www.loharchitects.com

Luis Felipe Infiesta Calzado/
SSCV Arquitectos
Sant Elies 11, despacho 99
08006 Barcelona
Spain
T. +34 93 202 1908
F. +34 93 202 1908
sscv-arquitectos@coac.net
www.luisfelipeinfiesta.com

Marc Dixon Architect
145 Russell Street
Melbourne, Victoria 3000
Australia
T. +61 3 9663 6818
F. +61 3 9663 6803
marcdixon@netspace.net.au

María Victoria Besonías, Guillermo de
Almeida, Luciano Kruk/BAK Arquitectos
arqbesonias@yahoo.com.ar
arqkruk@yahoo.com.ar
www.bakarquitectos.com.ar

Markus Wespi Jérôme de Meuron
Architekten
Caviano, Zurigo
6578 Caviano
Switzerland
T/F. +41 91 794 17 73
info@wespidemeuron.ch
www.wespidemeuron.ch

Masahiro Ikeda
1-20 Ohyama-cho Shibuya
Tokyo 151-0065
Japan
T. +81 3 5738 5564
F. +81 3 5738 5565
info@miascoltd.net
www.miascoltd.net

Mikan
5-49 Honcho Naka-ku
Yokohama 231-0005
Japan
yukou@mikan.co.jp
www.mikan.co.jp

Naoki Terada/Teradadesign Architects
2-19-13 ASK Building 2 F
Kabukicyo Shinjyuku, Tokyo
Japan
T. +81 3 6413 5700
F. +81 3 6413 5701
info@teradadesign.com
www.teradadesign.com

Paul Morgan Architects
Level 10, 221 Queen Street
Melbourne, Victoria 3000
Australia
T. +61 3 9600 3253
F. +61 3 9602 5673
office@paulmorganarchitects.com
www.paulmorganarchitects.com

Peanutz Architekten
T. +49 178 18 44 098
post@peanutz-architekten.de
www.peanutzarchitekten.de

Resolution: 4 Architecture
150 West 28th Street, Suite 1902
New York, NY 10001
USA
T. +1 212 675 9266
F. +1 212 206 0944
info@re4a.com
www.re4a.com

Rob Hootsmans/Hootsmans
Architectuurbureau
Johan van Hasseltweg 33 B
1021 KN Amsterdam
The Netherlands
T. +31 20 6360034
F. +31 20 6363532
info@hootsmans.com
www.hootsmans.com

Robert Nassar/Robert Nassar Design
315 West 39 Street, Suite 907
New York, NY 10018
USA
T. +1 212 629 8774
F. +1 212 629 8148
rsn2@earthlink.net
www.robertnassar.com

aaj Design
22 Bluff Road
andringham, Victoria
ustralia
+61 3 9598 1996
+61 3 9598 1997
aaj@bigpond.net.au
ww.saaj.com.au

aia Barbarese Topouzanov Architectes
39 Est Rue Saint-Paul
Montreal, Quebec H2Y 1H3
anada
+51 4 866 2085
+51 4 874 0233
ot@sbt.qc.ca
ww.sbt.qc.ca

ami Rintala
elgesensgate 82 A
563 Oslo
orway
+47 905 19005
ami@samirintala.com
ww.samirintala.com

hane Thompson/Bligh Voller Nield
65 St Paul's Terrace
ortitude Valley
ueensland 4006
ustralia
+61 7 3852 2525
+61 7 3852 2544
risbane@bvn.com.au
ww.bvn.com.au

HARE Architects
reitenfeldergasse 14/2 A
080 Vienna
ustria
/F. +43 1 817 78 30
ffice@share-arch.com
w.share-arch.com

Shubin & Donaldson Architects
3834 Willat Avenue
Culver City, CA 90232
USA
T. +1 310 204 0688
F. +1 310 559 0219
jtaylorpr@usa.net
www.shubinanddonaldson.com

Steven Haas
225 East Road
Alford, MA 01266
USA
T. +1 413 528 0625
F. +1 528 0684
haaseastrd@hughes.net
www.haas-tillinghast.com

Swatt Architects
5845 Doyle Street, Suite 104
Emeryville, CA 94608
USA
T. +1 510 985 9779
F. +1 510 985 0116
jfan@swattarchitects.com
www.swattarchitects.com

Takao Shiotsuka
301-4-1-24 Miyako-machi
Oita-shi, Oita 870-0034
Japan
T. +81 97 538 8828
F. +81 97 538 8829
shio-atl@shio-atl.com
www.shio-atl.com

Teeple Architects
5 Camden Street
Toronto, Ontario M5V 1V2
Canada
T. +41 6 598 0554
F. +41 6 598 1705
jlatto@teeplearch.com
www.teeplearch.com

Tom Kundig/Olson Sundberg Kundig Allen
Architects
159 South Jackson Street, Suite 600
Seattle, Washington 98104
USA
T. +1 206 624 5670
F. +1 206 624 3730
matt@oskaarchitects.com
www.oskaarchitects.com

Trevor Abramson, Douglas Teiger/
Abramson Teiger Architects
8924 Lindblade Street
Culver City, CA 90232
USA
T. +1 310 838 8998
F. +1 310 838 8332
sherry@abramsonteiger.com
www.abramsonteiger.com

UN Studio
PO Box 75381
1070 AJ Amsterdam
The Netherlands
T. +31 20 570 20 40
F. +31 20 570 20 41
k.murphy@unstudio.com
www.unstudio.com

Víctor Cañas
PO Box 340-2050
San José
Costa Rica
T. +506 253 2112
F. +506 228 1852
victor@canas.co.cr
www.victor.canas.co.cr